Understanding Industrial Practices in
Resistant Materials Technology

JET MAYOR

SERIES EDITOR: LESLEY CRESSWELL

Published in 2004 by:
Nelson Thornes Ltd
Delta Place
27 Bath Road
CHELTENHAM
GL53 7TH
United Kingdom

08 / 10 9 8 7

A catalogue record for this book is available from the British Library

ISBN 978 0 7487 9021 0

Edited by Judi Hunter
Illustrations by Jet Mayor and Peters and Zabransky
Page make-up by Northern Phototypesetting Co Ltd, Bolton

Printed and bound in Croatia by Zrinski

Acknowledgements
The publishers are grateful to the following for permission to reproduce
photographs and other copyright material:

British Standards Institution (p3); Department for Environment, Food and
Rural Affairs (Defra) (p22).

Every effort has been made to contact copyright holders. The publishers
apologise to anyone whose rights have been inadvertently overlooked,
and will be happy to rectify any errors or omissions.

Introduction

Understanding Industrial Practices in Resistant Materials Technology covers the key terms relating to industrial designing and manufacturing. It aims to support students undertaking GCSE Technology and GCSE Manufacturing (Double Award) courses. The book will help students and teachers understand what industrial practice is, what Systems and Control means and how to integrate these aspects into coursework using simple industry-related activities. For exam revision the book will be invaluable in helping students gain a good grasp of terminology and theory, both essential for exam success.

The first part of the book is set out in an A–Z format. Each photocopiable entry begins with a clear definition of the term, followed by an explanation and examples. Throughout the text there are frequent references to coursework requirements. This ensures students know what they have to do and why. These are followed by Coursework checkpoints which show the direct application of the term to students' own coursework.

The second part of the book provides photocopiable worksheets that are associated with the key A–Z terms. The worksheets and A–Z definitions can be used as teaching resources and support individual students in their own coursework.

What are industrial practices?

At KS4 the curriculum follows a design and manufacturing process that is based on an industrial model. In industry these designing and manufacturing processes enable manufacturers to make cost-effective products at a profit. At KS4 these activities are referred to as 'industrial practices'.

Many industrial practices make use of CAD/CAM and Information and Communication Technology (ICT). These have revolutionised the way industry works, enabling companies to communicate information quickly and to design and manufacture on a global scale. Developing this capability in schools has required access to specialist computer software and ICT links, to enhance designing, modelling, communicating, manufacturing and control.

Why do we need to include industrial practices in Technology and Manufacturing?

An understanding of industrial practices is part of the requirements of the KS4 curriculum for Design and Technology and Manufacturing. They are also practices used by industry to achieve the successful manufacture of innovative products that provide valuable exports. Industrial practices used in this way enable and encourage creativity.

We need to use industrial practices in the same way, as *processes* that enable students to design and make innovative products. We also need to provide a curriculum that is relevant and forward looking. Using up-to-date industrial practices will enable us to do this. Bringing manufacturing into the classroom will give students:

- a wider understanding of industry and what it has to offer
- an understanding of how ICT has revolutionised industry, enabling manufacturing on a global scale
- the opportunity for improved motivation and higher levels of achievement.

How to use this book

Industrial designers and responsible manufacturers are professional bodies of people dedicated to incorporating usefulness, good appearance, comfort, low cost, easy maintenance, safety and sales appeal in the products they design and make. Many factors are considered when designing products for sale, some are obvious, others are complex. Good designers have a grasp of most of these, even if they can't all be used or understood in detail. This book helps in an understanding of many of the issues, particularly those that could be examined in a GCSE Resistant Materials course.

The book addresses students directly in order to encourage independent learning. This means that many of the Coursework checkpoints and worksheet activities can be used for homework, or as focused practical tasks.

The A–Z terms are carefully cross-referenced with some terms appearing in a different typeface. This means that more information about that term can be found on another page. Students can look up the meaning of key industrial terms and use the definitions for exam revision.

The references to coursework requirements will help students understand why they need to include industrial practices in their own coursework. They can then use the Coursework checkpoint activities to help them undertake activities related to their own coursework.

The ready-to-use worksheets are also cross-referenced. Terms that are listed in the A–Z section appear in a different typeface. This provides a reference for the A–Z sheets that will need to be copied to support the worksheet activities.

aesthetic properties relate to the look of a product, matching its style with the **end-use**, e.g. matching aesthetic properties such as shape or form, **colour** co-ordination or contrast, or soft or hard surfaces for visual effect (for example, the combination of different hard and soft materials such as stainless steel and wood for a lamp). Aesthetics play a big part in attracting potential **consumers** to new products. Even when products function well they do not sell if they don't look good. To become familiar with what makes products look good it is an idea to closely observe natural shapes, colours and forms or look at the work of successful artists, **designers** and architects.

anthropometric data is information about body sizes and other features such as grip and vision limits. Many **designers** refer to tables of body sizes when designing products for people to use. Anthropometric data can be found in **British Standards** size charts and in some resistant materials books.

- Some products such as scissors and shoes come in a range of types and sizes.
- Car driver seats and watch straps are made to be adjustable to suit different sizes.
- Pushchairs tend to come in one size but are provided with adjustable straps.

Designers have to be careful that 'average' sizes are not used for all types of product. Imagine a door in a school classroom that was designed to be high enough for the average-sized student to walk through. For some people this would be far too low. Ouch! In this case **percentile** values are used to suit a percentage of the population who might use the product. Tables and chairs designed for very young children at school need to be quite different in size to those used by teenagers and adults.

architecture makes use of technology and materials for building exciting 3D structures. The forms these structures take can be an important source of inspiration for everyday products. This can range from the influence of Greek columns, Roman bridges and Gothic churches to geodesic space frame domes, spaceships or cranes. The shapes, styles and structures can be collected as pictures on a **mood board** and used as starting points for the design of much smaller products, such as clocks and furniture.

assembly is the process of putting a product together using separate parts or components. Each process in the product's assembly is carefully planned and written as a **work schedule**, matching the **materials** and equipment required with the assembly processes used.

Product **manufacturers** use assembly processes such as **joining** component parts and combining **sub-assemblies**. The assembly processes need to be:

- easy and fast, to make products at a profit
- cost-effective, making use of **standard components** where possible to reduce costs
- efficient, using waste management techniques to reduce waste.

In your coursework project you will need to:

- ■ use aesthetic judgements when designing your product, taking into account the values of your intended users.

In your coursework project you will need to:

- ■ use anthropometric data when designing your product, taking into account the **critical dimensions** of your intended users
- ■ record the actual body measurements needed for your product.

In your coursework project you will need to:

- ■ produce and use a detailed work schedule, which shows the order of assembly for making your product
- ■ choose standard components that are quick and easy to use
- ■ use minimum material allowances to reduce waste and cut costs
- ■ match materials and components with tools, equipment and processes, taking account of **critical dimensions** and **tolerances**
- ■ use assembly processes that enable accurate and rapid manufacture.

---

assembly line (production line) is where high volume products are manufactured or put together by teams of people, with each manufacturing process following another in a line. An assembly line is part of a production system to manufacture products quickly and efficiently.

automation enables the operation and control of production processes using electronic or computer control systems, such as programmable logic controllers (PLCs). Once automatic machines are set up, they repeat processes continuously without the need for further human control (unless something goes wrong). Automation requires scheduled maintenance procedures to make sure that machines run consistently. Although it uses fewer people for a given output it does require a skilled workforce to operate smoothly.

Most automated systems make use of sensors to monitor and control the machines. Sensors provide feedback to the system so that changes are made to the process when necessary. For example, sensors can detect faults or shut down a machine if there is a potential danger to the operator, the machine or the product.

Adaptable automatic computer numerically controlled (CNC) machines are the backbone of automation. They enable flexible, fast and accurate repeat manufacture and the production of cost-effective reliable products.

the **bar code** is a product identification code held on a computer. It can refer to a whole list of product information such as safety, cost, numbers in stock, date of manufacture, parts codes and product application. Many retailers now use the information to check stock levels and automatically re-order stock using Electronic Point of Sale (EPOS) till systems. Retailers also add their own information to the basic manufacturing bar code information to provide EPOS systems with retail prices. Workers on the tills then use a bar code reader to price products. Management also use the information to predict consumer trends and to determine or monitor overall profitability on any selected items.

In your coursework project you will need to:

- design a resistant materials product that could be manufactured in quantity

- choose the most suitable **production system** for your product and give reasons for choosing it

- understand what is meant by batch production, why it is used for manufacturing particular types of products and how it compares with other **methods of production**

batch production is used in industry for producing fixed quantities of a product, either for stock or to order. For example, highly skilled people working in teams could make a fixed quantity of 'designer' clocks in batches of 100. In this kind of batch production, each team member shares the responsibility for producing batches of high quality products. The whole product could have all parts made and assembled by each individual, or each part could be made by each team member and assembled afterwards. A time and motion study would determine the most efficient use of labour. Note that sometimes component parts are bought in if it is a cheaper alternative (perhaps the clock mechanism in this case). Highest profitability is achieved with high throughput, high sales and low stock levels.

Sometimes repeated batches can be made over an extended period of time to fulfil orders from retailers for specific product combinations or colours. This kind of production is called quick response manufacturing (QRM) and it has evolved through the increasing use of Information and Communication Technology (ICT). The retailer's

3

Electronic Point of Sale (EPOS) tills collect and send product sales information via electronic links to the **manufacturer** to enable a quick response to changes in demand. Batches of products in the required combinations or colours can then be delivered from stock or made to order in a very short time. Batch production used in this way, reduces the amount of stock that retailers need to keep, which saves space, cuts costs and delivers a better service to the **consumer**.

a **brand** is a name given to a product or process that is manufactured under a company's chosen brand name, **trade mark** or trade name. The brand name protects and promotes the product or process. This prevents it from being copied legitimately by competitors.

Historically, branding has been used at:

- trade level for **materials** or processes, e.g. at building industry trade fairs
- **consumer** level for finished products, e.g. 'Coca Cola', 'Sony' and 'Rolex' brands, which are promoted worldwide.

Manufacturers actively promote brand loyalty through lifestyle marketing to create consumer demand.

- In the sports and leisure industry, products are associated with competitive teams. Typical products include Rolex, Tag and Swatch watches, Slazenger rackets and Nike sports products.
- The brand name 'Hoover' became a word people used to name all vacuum cleaners and even the process of vacuuming. Dyson has tried to do the same thing with his vacuum cleaner and many people now refer to this cleaner as a 'Dyson'.
- Most people will buy the brand they know, or have heard of, rather than buy another similar product. The idea of **quality** is associated with each brand name. Try this yourself by collecting pictures of a range of similar products of similar cost and ask a friend to put them in order of quality.

the **British Standards Institution (BSI)** sets standards, testing procedures and quality assurance (QA) procedures to make sure that **manufacturers** make products that fulfil the safety and **quality** needs of their **consumers** and the environment. Manufacturers of aluminium ladders, for example, have to conform to established safety standards. The test procedures for checking safety have to comply with British Standards (BS) guidelines and must be carried out under controlled conditions. If a product meets the set standards a manufacturer can apply for the BSI 'Kite mark'. The product is tested and assessed at regular intervals.

Kite mark

- understand that using **standard components** can make manufacturing more cost-effective
- understand how **quality assurance (QA)** systems and **quality control (QC)** techniques are used to manufacture high quality products.

In your coursework project you will need to:

- take responsibility for recognising **hazards** in a range of manufactured products, manufacturing processes and work areas
- use safety information to help you assess risks when designing, making and using manufactured products
- work out and use test procedures to check the safety and fitness for purpose of your product, so it is safe for your intended user and the environment.

There are hundreds of British Standards relating to manufactured products. For example, these can include processes and tests that relate to:

- adhesives
- appearance
- construction
- durability
- ergonomics
- finish
- fitness for purpose
- machines
- maintenance
- material elasticity or spring
- material stability
- materials testing
- product life
- production processes
- safety
- strength
- toxic material (finishes, asbestos, etc.)
- wear.

The production process can also be tested to find the most suitable **method of production** and to prevent faults or problems occurring. For example, in product manufacturing, processes such as cutting or manufacturing parts and **assembly** systems undergo routine tests to make sure that quality standards are maintained.

Tests are made on a variety of end products ranging from toys to televisions.

cell or **cellular manufacture** refers to a manufacturing system that uses a small team of people to make and assemble a complex product. Each team member may complete several manufacturing processes before passing the product on to the next team member for further processes to be completed. This is called **teamwork**. It can make production work more interesting and can encourage a shared responsibility for producing high **quality** products as quickly as possible.

the **client** or **customer** is a person or organisation that asks for a product to be designed and manufactured. This is often the **manufacturer** or **retailer** of the product. The role of the client is to:

- identify a need or opportunity for manufacturing a product
- organise **key people** to get a product made and to the market
- agree a **design brief** with the **designer** or design team
- agree a development cost and raise money for the project
- agree a time-plan and deadlines for the project.

The client may also organise people, sales and distribution either personally or through a retail network.

colour is an essential characteristic of manufactured products and it is a powerful **marketing** tool that can encourage **consumers** to buy products.

Different colour combinations can change the atmosphere of an environment, creating stimulating or calming effects. For example, a different atmosphere is created if furniture or light fittings are painted, made of metal or natural wood. This applies to all furniture and

In your coursework project you will need to:

- understand that social, cultural, safety and **environmental** issues can have an effect on how colour is used for products.

fittings in different rooms in a home. Bright primary colours, or materials such as steel, glass, ceramics and plastics can look busy, efficient and clean. Soft pastel colours, neutral tones or natural materials such as wood, leather and many fabrics can look soft and relaxing.

Bright colours can help identify people at risk and those who can help them. This applies to ambulance, police and rescue workers, their vehicles and accessories and those people who work on motorways and building sites. **Safety wear** is nearly always in bright or reflective colours. Road safety signs and car warning triangles tend to be red, white and black for visual effect. Where **safety** is an issue many product types have to conform to specified colours. For example, the backlights of vehicles and machine safety 'off' switches are red, whereas controls used for starting machines tend to be green.

In different cultures colour is used in different ways. For example, in some cultures red is a symbol of peace and friendship, but in other cultures red symbolises danger, warning, revolution and war.

> **Coursework checkpoint:** *colour*
>
> ○ Take account of changing styles, the importance of colour in different people's lifestyles or the use of bright colours in safety wear when developing your product **design specification**.

composites are combinations of **materials** that are used together to provide enhanced **properties**. Some are manufactured using the latest technology and materials to fulfil specialist **functional** requirements. Industry uses composite materials where **performance requirements** are critical, or cost can be reduced. They are used for their strength, reinforcement or protection properties in specially-engineered, high-performance products such as racing cars, crash helmets, skis, tennis rackets and mountain bikes. Versatile composites, such as glass or carbon fibre and polyester resin, have regularly been used by industry for a variety of **constructions** ranging from small heat resistant products to seating, car bodies, boats, **prototype** product development and even small buildings.

computer-aided design (CAD) involves the use of computers to model product ideas in 2D or 3D and either view them on screen or print them out for review and reference. CAD **modelling** is a key part of the industrial design process because it enables **manufacturers** to test and modify ideas in 2D and simulate products on screen in 3D. These **virtual products** can be shown to **clients**, who can then choose which one they would like to see as a product **prototype**. This reduces the need for actually making a range of prototypes before final production, saving time and costs. One of the major uses of CAD for resistant materials products is for producing **working drawings** to support **manufacturing specifications** and production.

In your coursework project you will need to use:

- ■ a variety of graphic techniques to generate and develop design ideas
- ■ CAD to develop and model design ideas
- ■ CAD to produce accurate working drawings
- ■ CAD to develop component parts sizes and model costs.

▶

The use of CAD software (see worksheet 4 for a larger version of this diagram)

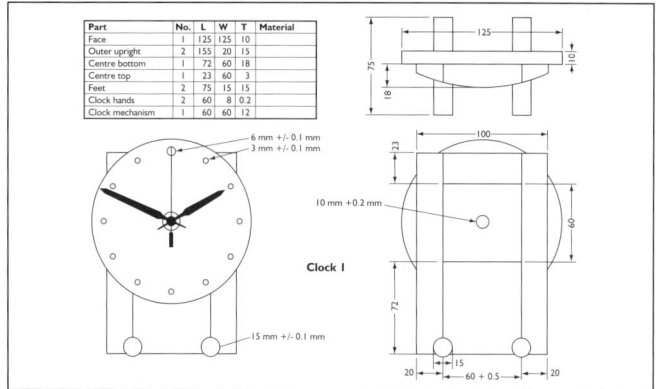

Designers and manufacturers can use CAD in various ways to:

- store style and **colour** information
- create, model and modify design ideas quickly and easily
- apply textures, renderings or shadings to drawings to create virtual products (wood grain, shiny or rough surfaces, lighting effects, etc.)
- present a 'virtual' 3D animation of a product using multimedia
- make accurate working drawings for manufacturing specifications
- produce general **assembly** drawings of individual component parts
- work out volumes and costs of the modelled parts
- produce 3D virtual products on screen or printed
- produce coded information of drawn component parts to drive **computer numerically controlled (CNC)** machines that can make them.

The use of **Information and Communication Technology (ICT)** enables CAD information to be sent electronically anywhere in the world between **clients**, designers and manufacturers, so that design decisions can be made quickly.

Coursework checkpoint 1: CAD

Use hand techniques such as drawing and sketching to generate first ideas quickly. Many designers still prefer to draw by hand and you will need to develop these techniques as well.

If you are short of ideas you can experiment in the following way:

○ Collect a range of architectural styles that interest you. Use these as starting points for style ideas.

○ Find an era or art style that interests you and use this as a starting point (Egyptian, cubism, art nouveau, pop art, etc.)

○ Try scanning collected designs into CAD software and then change the scale, proportion or colour.

○ Experiment with mirror images. Put a collage together on screen.

○ Stretch, shrink or change the shape, proportion or colour of available CAD images.

○ Design a logo or special lettering and print or cut it out so it can be applied to your product.

○ Use CAD for developing scale or full-size **templates**.

Understanding Industrial Practices: Resistant Materials Technology © Nelson Thornes 2004

5. computer-aided manufacture (CAM)

computer-aided manufacture (CAM) involves the use of computer systems to control manufacturing equipment, making it easier and quicker to produce cost-effective **one-off**, **batch produced** and **high volume** products. Once set up, the making process is very easy to repeat. CAM systems are used to cut or machine component parts but some CAM systems also use computers to control machines such as injection moulders, extruders and robotic **control systems**.

CAM automates production, repeats processes easily and precisely and is used in different areas of product manufacturing, such as:

● low level manufacturing aids, for example, computer generated printed **templates**

● plotter/cutter letters and logos

● one-off machining of specialised parts using fast, accurate CAM systems. For example, producing specialised tooling for machines.

● batch produced and high volume furniture manufacture, using CAM systems or **computer numerically controlled (CNC)** machines for efficient cutting, machining and surface **finishing** of parts

● high volume manufacture and **assembly** of many electrical product containers, such as mobile phones, kettles, radios, televisions, etc.

> **In your coursework project you will need to:**
>
> ■ understand how CAM is used in making one-off, batch and high volume products
>
> ■ use CAM to help manufacture your product, or part of it
>
> ■ use electronic or computer-aided machines for repeat processes such as printing, cutting, drilling or machining.

computer integrated manufacture (CIM) systems integrate the use of different functions of computers, including CAD/CAM, automatic storage and retrieval, robotic handling and material transfer, computer numerical control (CNC) machining and automatic inspection to enable fast, efficient and cost-effective manufacturing.

CIM systems reduce the product time to market and enable manufacturers to use just in time (JIT) methods of production and quick response manufacturing (QRM).

The target is to maintain a production flow that is able to cope with individual consumer preferences. Materials or parts are ordered and delivered to the factory on a JIT basis before being called for by an automated computerised system to be machined, checked and assembled by robotic systems and CNC machines to make a complete product. Each computerised workstation or cell completes its own dedicated task before the part or product is moved automatically to the next station. Many manufacturing plants now work on this basis including a number of high volume furniture manufacturers, domestic products manufacturers and most of the car manufacturing industry. The system allows consumers to state individual product preferences and have them included in the product manufacture. The whole system is extremely flexible yet can be controlled by a relatively small, but highly skilled, labour force, though it costs a lot to set up and maintain.

computer numerically controlled (CNC) machines enable manufacturers to make products accurately, economically and to increase efficiency. CNC machines are controlled by computer software. They enable automated manufacture of a very wide range of products in wood, metals and plastics. For example, CNC machines are used for the continuous production of plastic waste pipes, in high volume furniture manufacture, for high volume injection moulded plastic products and for cutting, machining and surface finishing parts. CNC machines include printers, plotter/cutters, lathes, routers, mills, lasers, injectors and extruders plus other specialist processing machines.

computer systems are used in industry for computer-aided design (CAD), production planning, data control, computer-aided manufacturing (CAM), computer integrated manufacturing (CIM) and Information and Communication Technology (ICT). The use of high-cost computer systems has revolutionised the design and manufacture of products, enabling just in time (JIT), quick response and global manufacturing.

Manufacturers use computer systems because they are adaptable, accurate and can work 24 hours a day, 7 days a week. They provide easy access to data storage and cost-effective, fast, high-quality production using computer numerically controlled (CNC) machines. The use of computer systems also tends to reduce the need for people to do tedious, repetitive work. This can lead to unemployment unless other types of jobs become available.

Manufacturing industry uses computers for general activities such as:

- administration
- managing product data, raw materials, components and stock using product data management (PDM) systems
- production planning and quality assurance (QA)
- electronic communications between companies and clients, using ICT
- marketing, costing and accounts.

Specialist computer systems are also used in manufacturing for:

- modelling ideas in 2D, 3D and virtual reality using CAD
- modelling ideas as CNC machined prototypes
- designing and producing working drawings using CAD
- planning sheet material layouts, to minimise waste
- CAM systems using CNC machines for production.

General and specialist computer systems can be represented by block diagrams that show how INPUTS are PROCESSED to achieve OUTPUTS.

▶

In your coursework project you will need to:

- understand how and why CNC machines are used to make **one-off**, **batch** and high volume products
- use CAM where appropriate to make part or the whole of your product.

In your coursework project you will need to understand:

- that computer systems are used in industry to support designing and control manufacture
- that systems are made up of inputs, processes and outputs.

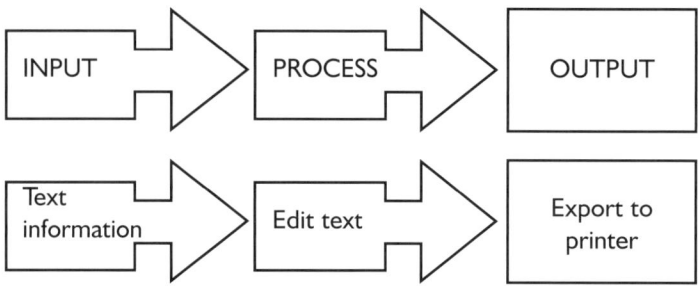

Block diagram to show an example of a general computer system activity

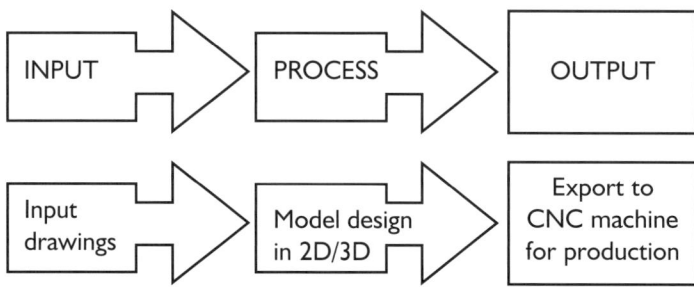

Block diagram to show an example of a specialist computer system activity

concurrent design and manufacture (CDM) is a system for organising and managing the development and manufacture of products. In this type of system all the research, **marketing**, design development, **production planning** and manufacture is managed by design and production teams who work closely together using **computer systems** to reduce the product time to market.

CDM is made possible by using **Information and Communication Technology (ICT)**, **product data management (PDM)** to handle and transfer data, **CAD/CAM** and **computer integrated manufacture (CIM)** systems to enable fast, efficient and cost-effective product design and manufacture. CDM makes it possible for **manufacturers** to use **just in time (JIT) methods of production** and organise **quick response manufacturing (QRM)** systems. In industry, the use of ICT and PDM systems enables design and production teams to have access to all the technical and organisational data about the product. Every time a change is made this data is available to all, so each can monitor and manage their section of product development. Where necessary, the information is also available to **clients** and **retailers** via a computer linked network.

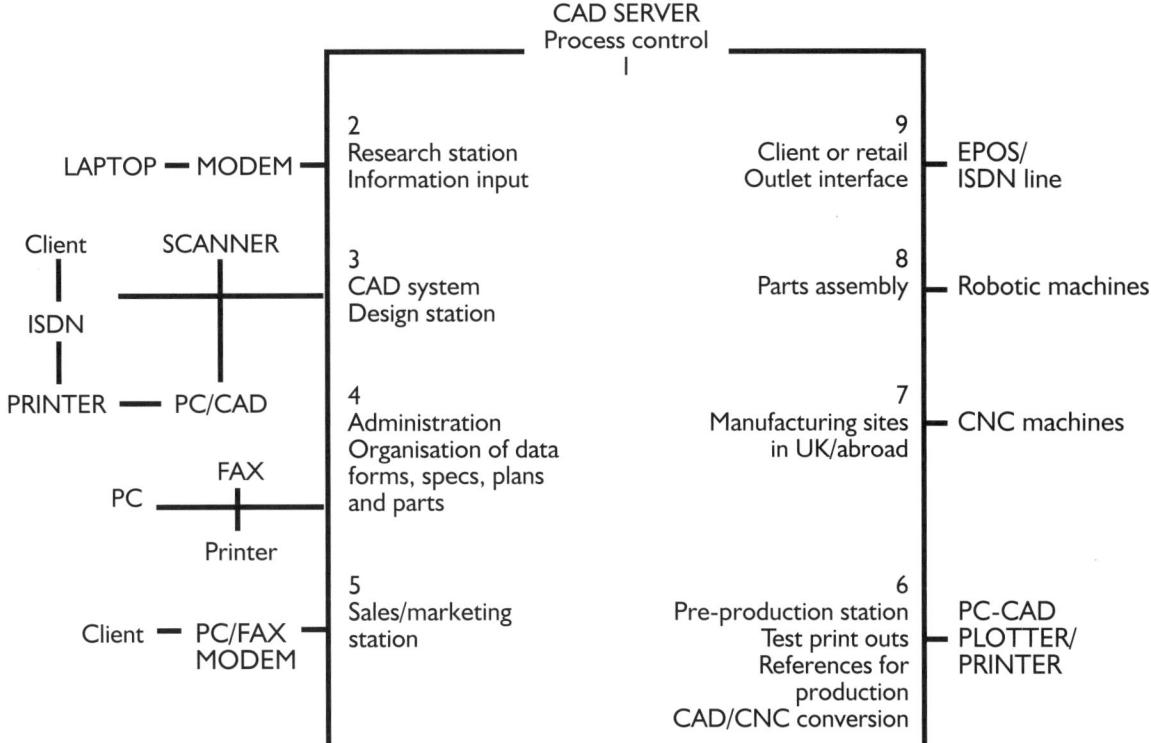

This diagram shows how concurrent design systems can integrate the use of different functions of computers in the design and manufacturing process

Coursework checkpoint: *concurrent design and manufacture (CDM)*

As a designer-maker you have to manage product design and development to make sure you finish on time with a high **quality** product, so you know how complicated it can be! You have the advantage that you don't generally need to communicate much information to other people during **one-off** manufacture.

If you take part in the **batch production** of a product you will need to:

- manage production as a team
- work out when you can use CAD/CAM and what for
- include all the technical and organisational information about the product in your production plan
- use a computer network to share information, so that any team member making a change can communicate this to others in the team. You might be able to do this on the Internet from home.

 the **construction** of a product affects how long it takes to make, how robust it is and how it looks. In industry, **designers** try to keep the number of parts down to an absolute minimum to reduce cost and improve reliability. However, this must not compromise style or **consumers** may not buy the product. Construction and appearance also depend on the **materials** used. For example, many metals can be joined by heat welding or be held together mechanically with nuts and bolts, rivets or screws. Some of the newer constructions use specialised adhesives. Many plastics can also be heat welded (polypropylene works well), joined mechanically or by adhesive but by far the greatest number of commercial plastic products are injection moulded or extruded. Wood is normally joined either mechanically with screws, pegs and knock-down fittings, or by using **joints** and adhesives such as PVA.

In your coursework project you will need to:

- understand the relationship between the **properties** of materials and manufacturing processes to make best use of them
- understand that combining or processing materials can give them more useful properties
- understand that mixing materials can give products more useful properties
- use materials that are suitable for product **end-use**.

consumers are the key people all products are aimed at. They buy the end product. Their needs are very important if **manufacturers** and **retailers** are to make a profit. A lot of effort goes into finding out what consumers really want or will buy and in **marketing** products. In many cases consumers and users are the same people, however, sometimes users are not the people who buy the product. An example is where parents who are consumers buy products for their children who are the users.

Consumers demand:

- high **quality** products that are fit for their intended purpose and meet consumer expectations
- stylish products with **aesthetic** appeal which reflect current lifestyle trends
- value for money products which last an acceptable lifetime (at least the guarantee period)
- products that are safe to use
- products that give pleasure in use
- products that meet all points of current consumer **legislation** related to the type of product being bought
- the exchange of defective goods for new ones or get their money back.

The consumer protection act gives ordinary people lots of rights and helps to guarantee that bought products are satisfactory for their intended purpose.

In your coursework project you will need to:

- choose the most suitable **method of production** for your product and give reasons for choosing it
- understand what is meant by continuous production, why it is used for manufacturing products and how it compares with other methods of production
- understand how **quality control** techniques are used to manufacture high quality products.

continuous production is used to manufacture very **high volume** products, such as extruded plastic pipes, injection moulded products and close tolerance engineering products such as machine screws. This type of production is highly automated and uses **machines** that can run continuously. Sensors monitor many continuous production machines, so that they stop automatically if a fault occurs. For example, the machines will stop if the **materials** overheat or parts are not ejected properly. Once the fault is corrected, the machines continue to function.

Continuous production machines use sensors to control the **quality** of the product as well as to monitor machine **safety**. This ensures the production of identical, high quality products, which range from simple washing-up bowls and plastic drain pipes to close tolerance engineering parts for products such as car engines, cassette mechanisms and watches.

control systems can be electrical, electronic, mechanical or computer controlled. Control systems are used in manufacturing to support all production **systems** including **one-off**, **batch**, **high volume** and **continuous**.

Control systems include:

- computer control (automation, CAD/CAM, computer numerical control (CNC), Information and Communication Technology (ICT))
- integrated manufacturing systems (computer integrated manufacture (CIM), product data management (PDM))
- quality control (QC) systems
- safety systems
- stock control systems.

Understanding Industrial Practices: Resistant Materials Technology © Nelson Thornes 2004

Control systems consist of a co-ordinated arrangement of activities working together, in which INPUTS are PROCESSED to achieve OUTPUTS.

In computer control, for example, the input could be the data from a **computer-aided design (CAD) working drawing**. The information is processed into code to drive a CNC machine such as a computerised milling cutter. The output is the final product that is cut by the CNC machine.

Some control systems are mechanical. Many older machines such as lathes and millers have lead screws that can be used to make the cutting tool or work piece travel automatically. These machines have no **feedback** systems to turn them off if something goes wrong. They rely on the operator to do this. This is an example of an open loop system as there is no automatic feedback to stop the machine.

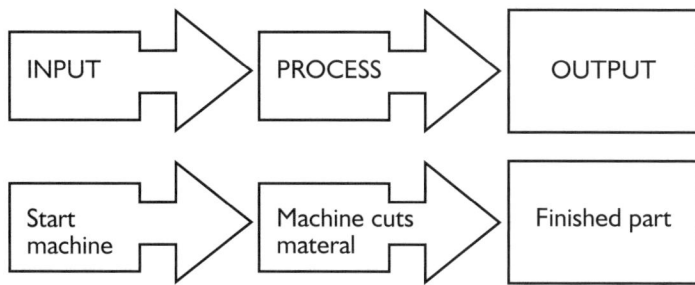

Block diagram of an open loop system showing computer control with no feedback

Some control systems incorporate feedback of information to make different processes function better or more safely. Most automated machines such as CNC lathes or CNC millers have end stops and door switches, so that if the tool travels to its set limit or the guard door is opened the machine stops.

A control system that incorporates feedback is called a closed loop system.

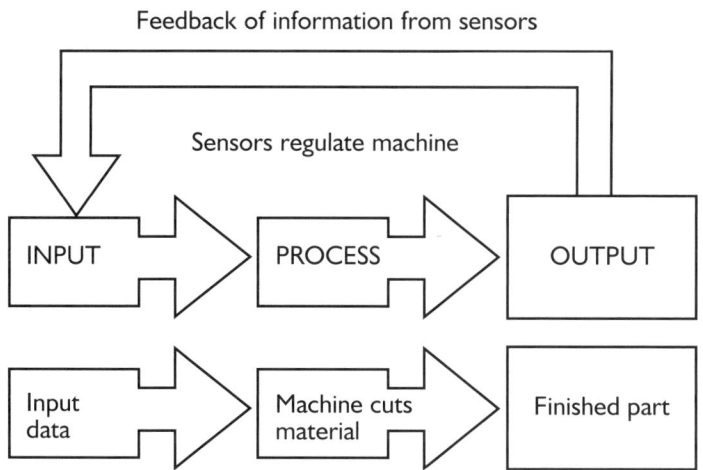

Block diagram of a closed loop system showing computer control with feedback

All closed loop systems use sensors to detect faults, or adjust to changing circumstances just as humans do. Electronic sensors can detect movement, heat, sound, light, position, vibration, etc. The control system in an automatic washing machine is an example of a

- understand that a control system is made up of inputs, processes and outputs

- design or model your product on screen using computer systems

- use computer controlled machines for easily repeatable processes such as machining parts or for cutting letters or shapes in vinyl

- use a quality control system that incorporates feedback so your product is fault-free

- use a safety system when manufacturing your product so your working environment is safe and the product is safe for the user.

closed loop feedback system. When the water level rises to the required level the system shuts the water off and begins to heat it.

Many control systems incorporate feedback of information to make the manufacturing process work well or more safely. For example:

- Materials handling systems, such as automatic fork-lift trucks, are used to move products and heavy tools around the factory, so that products are in the right place at the right time.

- Computer control systems are used to work out material costs, model ideas on screen, draft working drawings, work out efficient material cutting and for the automatic machining of parts.

- Computer control of processes includes heating (extruded pipes/casting), injection moulding, vacuum forming, shaping, cutting and polishing by machining.

- QC systems use feedback from checking routines to ensure the manufacture of high quality products of uniform consistency.

- Safety systems in manufacturing industry use feedback from sensors to stop machinery when faults occur or when machines are used incorrectly.

Control systems are used in manufacturing to:

- provide feedback that makes processes more reliable and safe
- repeat processes easily to make identical products of high quality
- maintain quality by producing accurate work
- speed up production
- automate tedious processes
- reduce waste.

In your coursework project you will need to:

- choose an easy and fast method of production so your product can be made cost-effective

- use computer software such as a spreadsheet to model the cost of the materials, components and labour.

costing is the process of producing an accurate price for a product that will make it saleable *and* create a profit. The level of costing depends on the **method of production**, which must be easy and fast so the labour costs are kept as low as possible.

- For example, domestic baths are made in acrylic by vacuum forming the basic shell and spraying glass-fibre and resin on the back to strengthen it. Although acrylic could be injection moulded, the size of a moulding machine and cost of tooling would far outweigh the profit made on even a large production run. A vacuum-forming tool is much cheaper to produce and can be quickly modified to cope with changing styles on lower production runs.

- In the case of a large production run for a mobile phone case, injection moulding is the best option. The cost of the machines and moulding tools can be written off over the predicted life cycle of the product. For example, if 100,000 mobile phone cases are made in a first production run and machine and tool costs are £100,000 then £1 is added to the final cost of each product.

Many other costs are involved such as design and development, **marketing**, procurement (getting **materials** into the factory), materials, labour, administration, overheads (buildings and energy), packaging, transport, distribution and sales.

Manufacturers have to cost their products very accurately and this can be quite complicated, so **computer systems** are being used more widely for estimating costs and forecasting profits.

Costing your product can be done in set stages and needs to include the following:

1 DIRECT COSTS such as materials and labour. These vary with the numbers of products made in a given period of time.

2 OVERHEAD COSTS such as rent, heat and electricity. These remain the same for one product or hundreds. They are often worked out as a set percentage of labour costs.

3 MANUFACTURING COSTS, made up of direct and overhead costs.

4 MANUFACTURING PROFIT, worked out as a set percentage of manufacturing costs.

5 The SELLING PRICE, made up of the manufacturing costs and the manufacturing profit.

Reducing costs:

The highest costs in manufacturing a small, batch-produced clock are the labour costs. The selling price could be lowered by:

- simplifying the design and manufacturing process to cut labour costs
- reducing the number of component parts
- using **jigs** or **templates** to speed up production and improve accuracy
- using **computer numerically controlled (CNC)** machines to speed up production and improve accuracy
- using less expensive materials and components.

the **creative** aspect of designing products is very important because product manufacturing is a competitive business. All **manufacturers** have to make their products attractive so that **consumers** want to buy them. **Designers** need to know consumers' preferences and what the **style** trends will be. They also need to have a good technical knowledge so that they can translate their creative ideas into products that can be manufactured easily at a profit.

Coursework checkpoint: *being creative*

Designing products can be creative, fun and exciting. As a designer you can enjoy experimenting with materials and styles but you also need to understand the properties of resistant materials so that you can exploit them. Try to keep the number of parts in any product to a minimum without compromising style. Look at the work of other designers to see how they use shapes and experimentation in their work. Consider what it is that attracts you to the products you own, or would like to own (**colour**, shape, texture, **size**, material, **quality**).

You can see the work of other designers by:

○ collecting ideas from newspapers, magazines, books and the Internet

○ going to shops, exhibitions and art galleries

○ looking at magazines, books and the Internet to find inspiration from other cultures.

In your coursework project you will need to:

■ understand that changing styles, the price consumers will pay for a product, **brand** image, hand or machine techniques and **environmental** concerns can influence the design of products

■ use the influence of other cultures, past and present techniques and styles, the work of other designers or recycled **materials** to inspire design ideas for new products

■ have a good understanding of how to work resistant materials so that you can exploit their **properties**.

a **critical control point** shows where quality control checks are made in the manufacturing process. Quality checks at critical control points ensure that the product meets the tolerances identified in a manufacturing specification, so the product is made accurately and to a high level of quality.

critical dimensions are the sizes of the parts of a product which must be made to specified tolerances to make sure all the parts fit together accurately. This is important if the safety or satisfactory working of the product relies on the fit of its different parts. For example, it is of little use making a drawer side 1 mm deeper than the hole it fits, though it may be possible to make it 1 mm less and the drawer still fits the hole.

critical path analysis enables designers and manufacturers to plan and monitor the critical path of products from design to the finished product. Critical path analysis is used to plan easy and cost-effective design and manufacture.

The critical path of a new product begins with the product design team working closely with the client to develop and test ideas. The design and manufacturing processes are then broken down into stages and put in order. Flow diagrams or Gantt charts are used to visualise the critical path of the product and to estimate the time needed to design and manufacture it.

Coursework checkpoint: *critical path analysis*

○ Plan a critical path for your next project, showing the deadlines for each stage of design and manufacture. You could produce a simple plan such as the one shown in the Gantt chart on page 17.

○ Some tasks may overlap, so that you are working on a number of tasks at the same time.

○ Try to estimate how long each task will take, so it fits the time available to you.

○ If you colour code your estimated deadlines and use a different colour to show your actual time taken you will be demonstrating the use of your time plan.

How one student used a Gantt chart to plan the design and manufacture of a sports accessory

Critical Path Analysis for sports accessory	WEEKS														
TASK	1	2	3	4	5	6	7	8	9	10	11	12–17	18	19	20
1 Brief/research/design specification	Δ ★	Δ ★	Δ ★												
2 Design ideas/testing/ feedback from users		Δ ★	Δ ★	Δ ★	Δ ★	★									
3 Modelling and prototyping/ feedback from users				Δ ★	Δ ★	Δ ★	Δ ★	★	★						
4 Production planning/quality control methods					Δ	Δ	Δ ★	Δ ★	Δ ★	★					
5 Manufacturing specification/ final costing						Δ	Δ	Δ	Δ ★	Δ ★	★				
6 Manufacture accessory										Δ	Δ ★	Δ ★	★		
7 Evaluation													Δ ★	Δ ★	Δ ★
8 Presentation to users and feedback													Δ ★		

Key: Δ = Estimated time
★ = Actual time

custom made (one-off, jobbing production) means designing and making a one-off product to a **client's** requirements.

a **design brief** sets out what is required by a **client** (customer). It needs to be simple and concise, explaining what needs to be done and have time limits. It should include relevant details but not the solution to the problem. The design brief is used to plan **market research** and to develop a product **design specification**. The client will often discuss the brief with the **designer** to make sure both are clear about what needs to be done. The client could be your teacher who sets the brief and the designer may be yourself.

Example of a design brief:
'A furniture manufacturer wants to develop a new range of attractive high **quality** clocks to meet the needs of young new-home buyers. The clocks need to be suitable for a modern kitchen environment and be in an attractive style. The finished working **prototypes** should be completed by 1 March. The designs should allow a first batch run of 50 to be made economically.' The proposed retail outlet is a chain of furniture stores.

In your coursework project you will need to:

- write, agree and use a design brief to develop a product
- use the design brief to plan research
- use the design brief to help develop a product design specification.

▶

The design brief explains:

- what needs to be designed (a type of product, such as a 'clock')
- what the product will be used for (its **end-use**; to tell the time and look good)
- who the product is for (the **target market group**, such as 'young new-home buyer')
- where the product will be sold (such as a specific shopping chain. This may help with the style.)
- where the product will be used (the environment: 'a kitchen'. This gives a clue to the sort of look and **finish** the product will need.)
- when the product prototypes must be finished ('by 1 March') and how many are to be made (50). This will affect the **production process**.

When you agree on a design brief and plan your research you need to consider how many products are required. Will the product be a **one-off** for a specific client or a product suitable for **batch** or **high volume production**? When you research your target market group you will also need to think about where the product might be sold. The intended retail market for your product will influence the style, quality and cost. Remember that it is a good idea to develop a reasonably simple product and finish it so you can meet the deadline for your coursework project.

Coursework checkpoints: *design brief*

1 In a Full Course GCSE in Design and Technology you may need to develop you own design brief based on a project outline given to you by your teacher. It is important to talk to your teacher to make sure that your design brief will enable you to cover all the coursework assessment criteria.

Example of a Full Course project outline:
You have been commissioned by a local craft shop to design a matching range of high quality candleholders that reflect a past designer style. The range should be suitable for table, wall and floor in a home environment. You are to make up one of each type in the range. You should be able to manufacture your designs in a batch quantity of at least 50.

2 In a Short Course GCSE in Design and Technology you are likely to be presented with a more focused project outline. It is important to talk to your teacher to make sure that your design brief will enable you to cover all of the coursework assessment criteria.

Example of a Short Course project outline:
You have been commissioned by a local craft shop to design a high quality candleholder based on the style of cylindrical tubes. The candleholder should be suitable for table, wall and floor in a home environment. You are to make up one candleholder. You should be able to manufacture your design in a batch quantity of at least 50.

3 In GCSE Manufacturing you will be given a design brief that tells you the client's requirements. The client could be your teacher or someone from industry.

Understanding Industrial Practices: Resistant Materials Technology © Nelson Thornes 2004

Example of a GCSE Manufacturing design brief:
'Candelabra', a local craft and lighting shop, is looking for a product to launch in its new store opening. The product should be suitable for display in the shop window. Initial market research has shown that holders for 6 squat candles are selling well in other craft stores. The first sample should be completed by 1 March.

a **design specification** outlines detailed criteria that guides a designer's thinking about what is to be designed. It sets out a list of specification criteria stating:

- the product's **end-use**, its function or purpose
- the needs and values of the product users – the **target market group**
- what the product should look like – its **aesthetic properties**
- how the product should perform so it meets the **quality** needs of the **client**, **manufacturer** and target market group (e.g. characteristics, **materials**, **components**, **finish**, **size**, weight, lifetime in normal use and maintenance)
- cost requirements of the product
- **safety** requirements of the product
- moral, social, cultural and environmental considerations
- the time limit for the product manufacture.

Coursework checkpoint: *design specification*

- Your product design specification can sometimes change slightly as you research your product.
- Remember to evaluate your design ideas against your product design specification criteria.
- Use your product design specification criteria and your final design proposal to help you develop your manufacturing specification. This provides detailed drawings and identifies the materials and components needed to manufacture the product.

In your coursework project you will need to:

- develop and use detailed design specification criteria
- use your design specification to generate design ideas
- **test** and evaluate your design ideas by comparing them to your design specification
- use your design specification and final design proposal to develop a **manufacturing specification**.

In GCSE Design and Technology the product design specification is developed from the design brief and research. It helps you generate and evaluate your design ideas, monitor the quality of your design work and develop a manufacturing specification.

In GCSE Manufacturing the product design specification helps you develop initial ideas of what the product will be like, how the product could be manufactured and how much it might cost. The product design specification needs to include criteria related to the performance requirements and cost of appropriate materials and components, the most cost-effective way to manufacture the product, quality and safety standards. You will need to use feedback from your client to help develop your final design proposal.

design themes are used by some **designers** to develop styling ideas for manufactured products. The theme can be based on artistic, cultural, social or environmental influences and is used as a starting point for design ideas. Examples of design themes include:

- ● Egyptian
- ● jungle
- ● geometric
- ● Greenpeace
- ● flash.

industrial **designers** need to incorporate usefulness, good appearance, comfort, low cost, easy maintenance, **safety** and sales appeal in a finished product. They work with other **key people** in the production team and may take on another role such as **marketing**, manufacture or sales.

The designer's role is to:

- ● agree an initial brief with the **client** (customer)
- ● check with users that an identified need is really wanted
- ● analyse existing products that may already meet the brief
- ● agree the **design specification** criteria with the client
- ● produce workable ideas based on the design specification
- ● consider environmental, social and moral implications of their product ideas
- ● take account of **legislation** and design safe products
- ● suggest **materials** and **production techniques**.

Product designers need to be aware of a wide range of techniques, processes and materials so they can design products that are fit for purpose. They need to reduce the number of working parts in a product and simplify production processes to keep costs down. Designers also need to be **style** conscious and have an eye for **consumer** trends.

As a product designer yourself, you have to take on a number of roles (see key people) though it is better if you can involve others in the design and manufacture of your product just as industry does.

designing for manufacture means designing a product that can be manufactured easily and quickly. To do this a **manufacturing specification** is drawn up. This guides the manufacturing process so that each product is made the same. The number of parts of a product should always be reduced to the minimum, as long as the product doesn't fail.

When designing for manufacture, you need to take into account:

- ● what the product should do
- ● the **scale of production**
- ● the availability, **properties** and cost of **materials**, components and equipment
- ● the manufacturing processes and ease of **assembly**
- ● the skills required to manufacture the product
- ● the expected cost and **quality** of the product
- ● how to make the product on time and at a reasonable cost.

development of ideas ready for manufacture is probably one of the most critical phases of the product design cycle. **Manufacturers** need to make a profit on the products they make so a number of product development issues are considered during **production planning** prior to manufacture. These include:

● finding different lower cost ways of making the product

● reducing cost and increasing reliability by reducing the number of product component parts, given that specification criteria are still met (this is called value engineering)

● packaging any moving or working parts in stylish containers or shells (like kettles and cars do)

● building models or **prototypes** based on **working drawings** to **test** the product for function and **style**.

disassembly is used to collect **market research** information about competitor's products. Some products can be disassembled quite easily, but if this is not practical it can be done by eye. Disassembly is used to analyse how products are constructed or how they work. Products are sometimes disassembled to analyse the **materials**, components and **finish** used, as well as their **quality** and value for money. The results of disassembly can be used to help with a product **design specification** and as a starting point for new ideas. Disassembling a product is a useful way of investigating **quality of design and manufacture**.

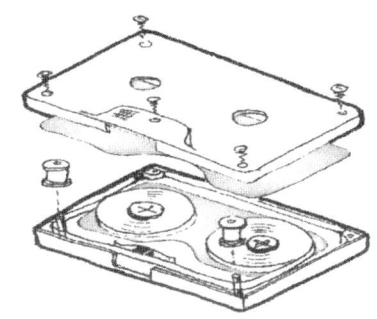

Parts: bottom + top case halves, screws, 2 spindles + tape guides, bottom + top low-friction liners, tape spools, central tape pressure pad

Cassette mechanism assembly

The logo of the Ecolabelling Scheme

ecolabelling is a system for monitoring the life cycle of a product from 'cradle to grave'. This is used as the basis for judging its environmental impact. The product's life cycle starts with the raw materials, continues through manufacture, distribution and use and ends with disposal (the grave). Many new products, especially those with a short life cycle, are made to be biodegradable. Examples include some supermarket polythene bags and milk containers.

The Ecolabelling Scheme uses **life cycle assessment (LCA)** to identify which parts of a product's life cycle harm the environment the most. It develops new criteria to reduce this impact. **Manufacturers** can apply for an ecolabel, have their products assessed and be licensed to use the ecolabel on their products.

Product Life Cycle					
Environmental impact	**Raw materials**	**Manufacture**	**Distribution and packaging**	**Use**	**Disposal**
Waste					
Soil pollution					
Water pollution					
Air pollution					
Noise					
Energy consumption					
Natural resources					
Effect on ecosystems					

Monitoring the life cycle of a product from cradle to grave

Electronic Point of Sale (EPOS) tills collect information about the sales of products from shops and stores. This data is sorted by computer and stock levels are calculated. Orders are sent to suppliers via electronic links for quick delivery of fast selling items. EPOS tills are used in **Information and Communication Technology (ICT)** systems, which enable **retailers** and **manufacturers** to respond quickly to **consumer** demand. If a product is selling very well, the manufacturer may have to **batch produce** more to fulfil the demand – this is called **quick response manufacturing (QRM)**. In some stores consumers can order particular manufactured products for delivery within a few days.

the **end-use** of a product is the purpose and target market for which it was made. Products are often described by their market sector or end-use; for example 'children's toys' or 'office furniture'. Designing with a specific end-use in mind makes the **designer's** job easier – the **performance requirements** and **design specification** of the product can be built around its function and who will use it.

energy can be categorised in two groups: renewable and non-renewable.

- Renewable energy including wave, wind, nuclear, solar power and vegetation comes from sources that can be replaced.
- Non-renewable energy comes from sources that cannot be replaced including coal, oil and natural gas. These are fossil fuels which were at one time living vegetation.

Legislation is making the development of renewable sources of energy more feasible and new wind farms and wave projects are beginning to appear. In warm climates, such as the Mediterranean, solar panels are often used to heat domestic water.

There are also many **environmental** groups interested in developing and farming natural sources of materials. For example, softwoods are farmed from managed forests in which a new tree is planted for every one that is cut down.

Although most plastic materials are currently made from non-renewable oil, new 'environmentally friendly' plastics such as Biopol have been developed. Biopol is made from renewable plant sources and will completely biodegrade at the end of its useful life.

environmental issues are increasingly important for the product manufacturing industry. **Manufacturers** are encouraged to use scarce resources economically and to manufacture products carefully, so that pollution to land, air or water is reduced. The **Ecolabelling** system is often used to monitor the environmental impact of manufactured products.

New technologies, such as biotechnology, are also being developed to manage environmental impact. Biotechnology can help reduce pollution caused by the disposal of products, such as those made from plastics. Some plastics materials, such as Biosystem™, are now designed to be biodegradable when composted. Some companies specialise in recycling plastic drinks bottles to make fibres or use plastic waste to make low cost sheet material for packaging products.

One way that manufacturing companies can help the environment is to reduce waste, which also helps them to cut costs. The three key approaches to raw materials waste management are:

- reduce
- reuse
- recycle.

Waste can end up in a landfill site or incinerator, so reducing waste saves money and helps the environment.

> **In your coursework project you will need to:**
>
> - consider the needs of the environment when developing your product **design specification**
> - use safe manufacturing processes and **materials** that reduce risks to the environment
> - take account of waste raw materials when designing and making.

Coursework checkpoint: *environmental issues*

○ Follow **safety** regulations and instructions so you dispose of waste products safely.

○ Use waste management principles in your product development. For example, if you are using sheet materials an efficient layout plan will help. Ask yourself these questions to help the environment:

Can you design and make more efficiently using fewer parts or less material?

If you can't prevent waste, can you reuse it?

If you can't reuse waste, can you recycle it?

○ Don't forget that waste needs to be added to the cost of your product.

---

ergonomics is about trying to match the design of products with human body shapes and sizes. Designers use anthropometric data to make products as easy as possible to use by the target market group.

- most kitchen work surfaces are at a height suitable to work at whilst standing. Dining table heights are lower because they are made for sitting at

- scissors are made with different handles for left and right-handed people to improve grip and cutting control

- the controls in a car need to be within easy reach and also be easy to understand by sight and touch. Controls, such as the steering wheel and brakes, have to be easy enough to operate by the weakest or smallest driver.

- machines in work areas should be easy to reach and operate as well as safe to use

- colour plays an important part in the design of controls such as stop and start switches and warning lights.

the **European Standards Organisation** sets performance standards for products used in Personal Protective Equipment (PPE).

PPE can be used for protection in the workplace or in high-risk activities, such as motor racing. All protective equipment marketed in Europe has to fulfil specific performance requirements and quality standards so it can carry the European Standard CE mark. It is illegal in Europe to sell any PPE item unless it carries the CE mark.

In your coursework project you will need to:

- recognise the benefits of particular fasteners, fixings or adhesives for particular **materials**

- pay attention to any **health and safety** requirements when using fasteners, fixings or adhesives (for example when making toys for children or using solvent or rapid set adhesives).

fasteners, fixings and adhesives are used as standard components. Industry generally uses the lowest cost fastening method possible, as long as it lasts the predicted lifetime of the product.

- Non-permanent fasteners include screws, nuts and bolts. A wide range of knock-down fittings such as 'bloc-joints' and 'scan-screws' are used for flat-pack furniture. Panel pins are used to temporarily fix thin sheet material whilst it is being glued.

- Permanent fasteners include rivets (solid and 'pop or blind'), nails, staples, pegs and dowels.

- Fixings include shelf brackets, hinges, hooks, handles and knobs, locks, runners, wheels, slides and castors. Industry uses the lowest cost fixings available, as long as they meet product design specifications and target market needs.

- Adhesives like PVA wood glue soak into the wood to bond the surfaces together. Araldite works by chemical action to bond metals together. Solvent adhesives work by dissolving the surface of the material rather like a cold weld. They are used on many plastics materials such as PVC, styrene and acrylic (Tensol cement is used for acrylic). Nylon and polythene cannot be easily solvent welded but are heat welded. This is how most polythene bags are sealed.

Understanding Industrial Practices: Resistant Materials Technology © Nelson Thornes 2004

feedback is when information is fed back into a **system** to control and improve the progress of the system. For example, most **computer numerically controlled (CNC)** or automated **machines** use feedback from sensors that stop the machine if the tool reaches a limit switch or end stop. Sensors are also used to check if tools get jammed or a motor stalls for any reason. An automatic washing machine has feedback sensors to detect things such as water level and water temperature. When the set limits are reached the next part of the process continues. When a system incorporates feedback it is called a closed loop system.

finishing is a **quality assurance (QA)** procedure to make sure that the product is fault-free, clean and matches the **manufacturing specification**. **Manufacturers** use finishing processes to improve surface **quality**, so it is suitable for its **end-use**.

Finishing is used to:

- make component parts fit accurately
- prepare a surface ready to take a protective coating or finish, such as sanding a wooden construction smooth before applying a wax or paint sealant.

Industrial finishing processes can be physical (mechanical) or chemical.

- Physical finishing processes include machining, sanding and buffing, depending on the **materials** being used.
- Chemical finishing processes include etching copper, anodising aluminium, plating steel with zinc or priming with a sealant such as resin or paint. These processes can cause environmental damage.

Finishing can be **aesthetic** for appearance and feel or **functional** to give a performance property, such as water-resistance or non-slip.

- Wooden products need surface finishing to protect them from dirt, damp and pests. Wood can be finished using cellulose, paint, wax, micro-porous stain or varnish (lacquer or polyurethane).
- Ferrous metals such as steel need surface finishing to reduce corrosion. Stainless steel, aluminium, zinc and brass are corrosion resistant. Metal can be painted, powder-coated, waxed/oiled, anodised (chemically treated), etched or plated. For example, plating steel with chrome or zinc makes it corrosion resistant and gives it an attractive appearance.
- Plastics are generally self-finished, painted, buffed or polished.

In your coursework project you will need to:

- incorporate feedback into your own design and **manufacturing system**
- use feedback from your **client**, **target market** or user to help you modify and improve your design ideas.

In your coursework project you will need to:

- understand that finishing processes can improve the quality of the product
- use finishing processes that can give your product enhanced aesthetic or functional properties
- understand that some finishing processes are not environmentally friendly.

Coursework checkpoint: *finishing*

- ○ Build product finishing processes into your quality assurance system.
- ○ Check all component parts are secure and work properly before applying a finish.
- ○ Inspect the product for surface and mechanical faults.
- ○ Make sure the materials you have used are suitably smooth before applying a finish.
- ○ Remember that the surfaces of some resistant materials are self-coloured and simply need buffing and polishing (most plastics and some metals such as stainless steel, brass and aluminium).
- ○ Remember that the quality of finish will attract potential **consumers**. Surface preparation is therefore very important. Check all sealed or painted finishes for dull patches or rough surface areas.

---

fitness for purpose is about making sure that the product is suitable for its intended **target market** and **end-use**. A product that is fit for its purpose is well designed and well made so its **quality** meets the needs of the **specification** and end-user.

The product may have to meet certain **safety** standards, which is important with products such as children's toys. The fitness for purpose of a product is evaluated against all the specifications and by performance, price and **aesthetic** appeal. The quality needs of the user are influenced by the product's:

- performance in use – is the product robust enough to do its job well and safe to use? Is the product easily maintained?

- value for money – can the product be sold at an attractive price? Does it give added value?

- appearance and the image it gives the user – is the product well designed and attractive to the target market?

a **flow diagram** is used for planning and working out the **stages of production** of a manufactured product. **Manufacturers** often use flow diagrams for **critical path analysis**, to plan easy and cost-effective manufacture. Flow diagrams can be used to plan where and how the **quality** of a product is checked.

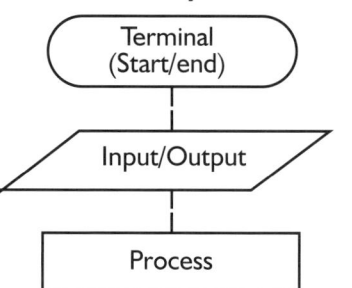

Standard symbols

Terminal (Start/end)

Input/Output

Process

Decision

Coursework checkpoint: *flow diagram*

○ Include a flow diagram in your production plan to show where and how you will check for quality.

○ Using a flow diagram with feedback loops will demonstrate your use of **quality control**.

flow-line production is used for manufacturing large quantities of products for stock or to order. The system involves a **continuous production** line where parts are made and **assembled** in a logical order. A number of workstations make or assemble complete **sub-assemblies** and then pass these to the next stage of manufacture. Most **high volume** products are made this way including cars, personal hi-fis, televisions, cutlery, toys, etc.

Understanding Industrial Practices: Resistant Materials Technology © Nelson Thornes 2004

functional properties relate to the performance of a product, matching these with **specifications** and the **end-use**.

Functional properties can relate to:

- the way a **material** performs, such as its strength, elasticity, surface qualities (slippery, etc.), heat resistance, weatherproof, weight
- the way a product performs, such as being durable, delicate, robust, accurate.

For example, the functional requirements of a space helmet could be strong, ultra-violet resistant, lightweight, airtight and heatproof. The functional requirements of a child's wooden toy car could be durable, non-toxic and robust.

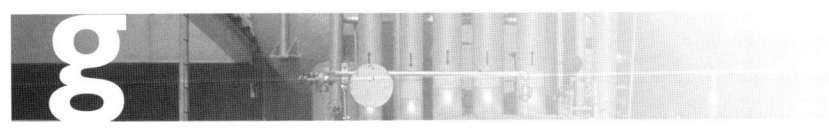

a **Gantt chart** can be used to plan a project, showing the tasks to be done and the time available to do them. You could plan your project in the following way:

- List the tasks in the order in which they need to be started.
- Use a horizontal line to show how long each task will take.
- Plan the tasks that might overlap and those that can be done at the same time.

In your coursework project you will need to:

- produce and use a plan to show the stage deadlines for the design and manufacture of your product
- use a time plan to show how long each stage of manufacture will take
- work to realistic deadlines.

Task	Weeks											
	1	2	3	4	5	6	7	8	9	10	11	12
1												
2												
3												
4												
5												
6												
7												

A Gantt chart

Coursework checkpoint: *Gantt chart*

- ○ You can use a Gantt chart to help plan your next project.
- ○ Find out the number of weeks available and the deadline for your project.
- ○ Copy the layout of the chart shown above and list the tasks you have to do in order, one after the other (see worksheet 25).
- ○ Use a coloured horizontal line to estimate the number of weeks each task will take.

○ Remember to plan which tasks you can do at the same time – for example, you may be able to develop your **design brief**, undertake research and develop your **design specification** at the same time (concurrently).

○ Use a different coloured horizontal line to show the actual time each task took. If you do this you will show where you had to make any changes to your production planning (see the Gantt chart on page 17).

○ Remember to record and explain any changes you make to the design or manufacture of your product resulting from problems in meeting your coursework deadline. This will enable you to explain how you could manufacture identical products in **high volume**.

a **generic name** is a common or general name for a material, such as 'wood', 'metal' or 'plastic'. Products need to be made from specific materials such as 'chipboard', 'aluminium' or 'acrylic'. The names of materials are sometimes taken one step further by manufacturers, who use brand names for materials designed and made by themselves. For example, a brand name for acrylic is Perspex™, which is made by ICI.

global manufacturing is where a product is researched, designed, manufactured and marketed in different locations. For example, a product can be designed in the UK and manufactured in another country such as China, where the manufacturing costs may be lower. Global manufacturing is possible through the use of Information and Communication Technology (ICT), which enables designers and manufacturers to exchange product information via electronic links. Global manufacturing can affect employment in different countries.

grain refers to the structure of a basic material. In wood, growth is a seasonal process, shown by the pattern of the annual rings, which give it a characteristic appearance. As the tree grows the wood tissue also grows to form long tube-like cells, varying in shape and size. These cells grow parallel to the length of the trunk to give wood its characteristic grain. However, even materials like metals and plastics have grain that is dependant on how the original material was formed. If a plastics part is injection moulded then the grain is the direction of flow of the plastic into the mould.

Softwood such as spruce is used for boat masts because its long cells make it long-grained. This makes it flexible so it resists breaking when bent. Hardwood is slow growing and has a close grained structure that makes it strong and tough.

a **hazard** is a source of potential harm or damage or a situation where an accident could occur. For example, chemical splashes are a hazard because they can cause skin damage. To reduce the risk of the hazard, chemicals need to be handled with care.

In your coursework project you will need to:

■ consider the shape of your product parts to make the best use of materials

■ take into account the direction of the material grain for greatest strength or effect.

In your coursework project you will need to:

■ take responsibility for recognising hazards in **materials** and manufacturing processes.

20

hazard analysis means identifying all the potential **hazards** and risks involved in manufacturing, using and disposing of products. Manufacturers use **risk assessment** to assess all risks to people, the environment and in manufacturing processes. Every source of danger must be removed or identified. **Safety** procedures have to be written down so that industrial accidents can be avoided. The table below shows potential hazards and safety procedures used in work areas.

In your coursework project you will need to:

■ take responsibility for recognising hazards in manufacturing processes and work areas

■ follow safety rules to reduce risk when designing and making your product

■ carry out safety checks on tools, equipment and machines

■ work safely with **materials** and equipment when manufacturing your product

Potential hazards	Safety procedures
Injury from: ● materials (hot, sharp, when transporting). ● hot areas (irons, glue-guns, hot air guns, hot metal).	● Wear safety gloves. ● Handle all materials with respect and care. ● NEVER hold materials on to machines using an apron or loose body clothing.
Injury to eyes, hair, fingers, mouth, body from: ● drills, routers, mills, lathes, buffers, sanders. ● saws and guillotines.	● Check that STOP switches work properly before work starts. ● Set and check tool guards. Clamp smaller pieces. ● Wear safety glasses, goggles and protective clothing. ● Wear a face mask when using buffers and sanders. ● Tie back loose hair and clothing.
Injury from: ● chemicals (varnishes, sealants, cleaners and adhesives). ● hot areas (irons, glue-guns, hot air guns, hot metal).	● Read instructions before using chemicals. ● Wear safety clothing especially for eyes, hands and feet. ● Dispose of chemicals safely after use.
Injury from: ● the working environment.	● Check for hazard warning signs near machines. ● Check work surfaces and floor are clean and not slippery. ● Check ventilation is adequate. ● Check for trailing cables, sharp corners, overhanging workpieces or open vices. ● Switch machines off and put tools away after use. Clean up spills. ● Know where fire fighting equipment is kept. ● Be aware of others and know basic first aid procedures.

Some of the potential hazards to look for in work areas

health and safety (H&S) at work is the responsibility of employers and employees. The Health and Safety at Work Act 1974 requires manufacturers to follow strict rules and regulations and to have an H&S system in place. They must undertake a risk assessment to make the workplace as safe as possible. Employees are required to follow safety procedures to reduce risks when using materials, machinery and manufacturing processes.

Manufacturers are also required to ensure a product's safety so that no harm can come to the user or the environment during normal use. For example, it is illegal to sell any item of protective wear in Europe unless the product has the CE label. If the product is tested and fails to meet the appropriate safety standard it is withdrawn from sale and the manufacturer is prosecuted.

Coursework checkpoint: *health and safety*

- ○ Know about the possible **hazards** and the relevant safety procedures in your working environment.
- ○ Read the safety rules in your working areas.
- ○ Keep your work area tidy.
- ○ Handle tools and equipment with care.
- ○ Check that **machine** guards are in position.
- ○ Read instructions and labels or ask for help when using potentially hazardous equipment, materials and chemicals.

high volume production (mass production) is used for manufacturing large quantities of manufactured products, for stock or to order. High volume production is a cost-effective method of making identical products for specific market sectors, such as children's toys or personal hi-fis. A target number is decided on for the production run and this dictates how long the run will last. At the end of the run a decision is made whether to continue or move to a different product.

This kind of production is planned for ease-of-manufacture, and where possible uses modular or standardised materials, components, equipment and manufacturing processes. Quality control checks throughout manufacture ensure that each product is identical. High volume production is used to make a wide range of products such as cutlery, CDs, hi-fis, toys, cars, televisions, computers, plastic drainpipe, and a whole host of injection moulded plastics products. Some of these products don't change quickly with fashion and any changes that are made need to be easy to implement. Some of these products are produced by continuous production methods but they could also be batch produced. The chosen method of production depends on numbers and the demands of the market.

21

22 **industrial practice** is about planning, designing, making and selling products in a commercial market for a profit.

To sell their products, **manufacturers** have to make judgements about the **target market** and provide products that are needed. The processes manufacturers use include activities such as **market research**, product analysis, working to a **design specification**, generating ideas, using **computer-aided design/computer-aided manufacture (CAD/CAM)** for production and using **quality control** to ensure accuracy and reliability. Product development and manufacture is a high cost, high risk business.

Several **key people** are involved in the design and manufacture of products, including **clients**, **designers**, manufacturers and **consumers**. Other people may be involved such as market researchers, engineers and **retailers**.

The industrial practices used by manufacturers depend on the type of products they make and the consumers they are selling to. For example, manufacturers of mobile phones may use industrial practices such as:

1 Design activities
- Develop a design brief.
- Product analysis – of similar commercial products.
- Market research – the target market, **style**, **colour**, trends, values issues, new technologies.
- Internet research – planned and focused.
- Use a design specification to generate and evaluate ideas.
- Use a **design theme** or **mood board**.
- Design **quality** and **safety** into the product.
- Work to a price range.
- Present ideas to the client/customer and use **feedback** to improve the design proposal.
- Use CAD for **modelling** and **prototyping**.
- Use CAD for **working drawings**.

2 Manufacturing activities
- **Test** a **prototype** product before production.
- Use CAD generated working drawings.
- Develop a **manufacturing specification**.
- Produce a **materials** and parts list to cost the product.
- Plan production in **batch** or **high volume**.
- Choose efficient processes for easy **assembly**.
- Check that sufficient technical expertise and equipment is available.
- Use **systems** to control production using **jigs**, **templates** and/or **computer numerical control (CNC)**.
- Use quality control.
- Use safe working practices.
- Test and evaluate against **specifications**.
- Evaluate product against target market requirements.
- Use feedback to suggest improvements.

In your coursework project you will need to:

- understand the key roles of the client, designer, manufacturer and consumer

- understand and use industrial practices, and systems and control when designing and making your product

- choose an easy and fast method of production so your product is cost-effective to manufacture.

▶

In your coursework project you will need to use:

■ CAD to develop and model design proposals

■ CAD to produce accurate **working drawings**

■ CAD to develop layout or parts plans

■ computer software to present work, produce specification sheets and model costs

■ **flow diagrams** to guide production and control quality

■ CAM in single item production or in **batch production**.

the use of **Information and Communication Technology (ICT)** has revolutionised the way the manufacturing industry works, enabling companies to communicate information quickly, and to design and manufacture on a global scale. ICT makes use of electronic and **computer systems**, which can:

1 Enable business partners to 'talk' to each other via electronic links such as:

- Integrated Services Data Network (ISDN)
- the Internet
- Electronic Point of Sale (EPOS) tills
- product data management (PDM) systems.

2 Provide computer control such as:

- computer-aided design (CAD)
- computer-aided manufacturing (CAM) using computer numerical control (CNC)
- computer integrated manufacturing (CIM) systems.

Electronic links can be used to:

- link business partners by e-mail or video conferencing
- import **style** information from across the world
- source and handle data such as ordering stock
- send CAD information to **clients** for approval
- enable a client to see **virtual products** on screen
- send a **manufacturing specification** to a distant production site
- design in one location and manufacture in another (**global manufacturing**)
- enable **just in time** (JIT) manufacturing
- monitor **quality** on a production line using sensors and/or digital cameras
- provide sales information through EPOS terminals
- enable **quick response manufacturing** (QRM)
- link **retailers** and **manufacturers**.

Understanding Industrial Practices: Resistant Materials Technology © Nelson Thornes 2004

- E-mail (to share information with other students or to contact manufacturers)
- The Internet (to find out information about materials, processes and products)
- Video conferencing (to share information with other students).

2 ICT systems enable the use of CAD/CAM for product design and manufacture. Investigate how you could use computer software to help you:
- analyse research information
- scan in style information
- produce a **design specification** sheet
- **colour** your product in different ways on screen
- record your product development using a digital camera
- produce working drawings of the product and its parts
- design a manufacturing specification sheet
- design a table or flow diagram to plan the critical path of your product
- design, print and cut a paint stencil or self-adhesive vinyl logo for your product
- export design information to a CAM machine (router, mill, lathe, vinyl cutter) to make a part of your product.

inspection is the quality control (QC) process used to check if a product or part matches its specification. Inspection takes place as each component part is manufactured, checking for criteria such as size, dimensional accuracy, **tolerance**, **colour** match or surface **finish**. It may also take place after **assembly** to **test** for fit and final surface quality.

interiors are designed to have **aesthetic** and **functional** properties in domestic, business or public buildings. They need to look attractive but have to fulfil functional **end-uses**, such as being relaxing or efficient. Interiors are full of manufactured products made of wood, metal and plastics. Looking critically at interiors can give you lots of ideas for manufactured products that could be made in your school workshops. You will need to ask yourself what each product does, or needs to do, how it works and how it could be improved in some way.

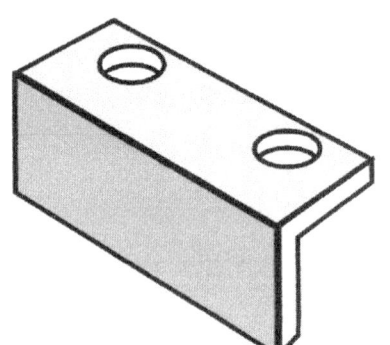

Diagram showing a simple workshop jig

jigs and templates are used in industry to speed up production and increase accuracy. A jig is a device used to hold a piece of work so that a manufacturing process can be repeated easily and accurately. Jigs do not need to be complicated to work well and you can make very simple jigs for your own work to help with production. For example, you could:

● drill a piece of steel angle bracket and use it as a guide or jig for drilling dowel joint holes in the same place in a wooden construction

● cut or sand pieces to the same length by making a very simple wooden block with an end stop and a slot for a saw.

templates are flat cut-out shapes that are used to copy the same shape accurately as many times as required. They can be used to mark out, check or test a finished piece for size, outline shape and location, such as the centres of holes for drilling. In industry templates are normally produced from computer generated diagrams. If a template is to be used many times it is made from a rigid material such as thin wood, plastic or metal. Paper or card is used for templates that are only used once or twice.

> **Coursework checkpoint:** *jigs and templates*
>
> ○ Identify jigs that could be used in the workshop.
>
> ○ Design a template or jig of your own to enable a repeat process and increase the accuracy of your production.

jobbing production or custom made production, means designing and making a **one-off** product to a **client's** requirements.

the **joining** techniques used for **materials** depends on the type of material to be joined, the equipment available and the **end-use** of the product. All joints are designed to increase the surface area of the material at the joint for greatest strength. The most common joining techniques are mechanical, fusion or heat welded joints and using adhesives, but whichever technique is used, **quality** and a professional-looking result is essential. Industry tries to choose the lowest cost method that does the job effectively for the predicted lifetime of the product.

1 Mechanical joints are used when the materials to be joined:

● are difficult to hold together by other methods

● need to be adjustable or flexible in use

● need to be accessible (such as the battery compartment of a torch).

Mechanical joints include those that use nuts, bolts, screws, rivets, pegs, pins, hinges, knock-down fittings, friction fittings or specially cut shapes in material such as dovetails. Nearly all joints in wood are mechanical, though laminating with adhesives is possible. Some plastics products use mechanical fixings but most are moulded to a 3D shape during production or clip together.

2 Fused or welded joints are used on metals and plastics when a permanent bond of great strength is required.

- Welding is a permanent joint made by heating the material to its melting point, at the joint. Most thermo-plastics can be heat welded which is used for large fabrications such as polypropylene water tanks.

- Fusion is when a joint is made using heat to melt a bonding material. For example, a brazing rod or silver solder can be used to fuse steel at much lower temperatures than by welding.

- Thermo-setting plastics and all types of wood cannot be joined by welding or fusion.

3 Adhesives are suitable for nearly all resistant materials. They work by surface penetration (soaking), solvent welding (dissolving) and chemical action (epoxy resins). All adhesives need to be used with care, as they can be toxic or harmful.

4 Wood joint systems include: mortises; laps; biscuit; comb; dovetail; dowels or pegs; laminating; screws; nuts and bolts. Most joints in wood are reinforced with an adhesive for strength and stability and are designed to increase the surface area of the material at the joint for greatest strength. For example, a mortise and tenon joint is far stronger than a dowel joint so it is still used for making outside wooden doors.

Some of the newer industrial jointing systems have developed because of the need to speed up production using machines. Biscuit, peg joints and knock-down fittings are mainly used for joining flat sheet materials such as MDF flat-pack constructions. They are increasingly being used on other types of furniture such as wooden tables and chairs.

5 Metal joint systems include: nuts and bolts; set screws; self-tap screws; rivets; welding; brazing; and soft-soldering.

6 Plastics joint systems include nearly all those used for metal, including friction fit. Not all plastics can be solvent welded. This process is used on acrylic sheet and products like PVC pipe joints, though many of these now use mechanical, watertight friction fittings. Welding is used for specialist applications like low cost polythene freezer bags. These are manufactured as a continuous tube by extrusion. A strip heater then seals the tube at regular intervals before it is perforated and rolled, ready to be used as tear-off bags.

25 **just in time (JIT)** manufacturing makes use of Information and Communication Technology (ICT) to help plan the ordering of materials and components, so they arrive at a factory just in time for production. This is a complex system that requires careful planning between a manufacturer and its suppliers. JIT is often used in quick response manufacturing (QRM), where goods are produced quickly in the exact quantities needed to fulfil consumer demand. JIT is used in combination with QRM because it:

- reduces the need for stockpiles of raw materials waiting to be used

- reduces the space needed for keeping raw materials in stock

- reduces the levels of finished goods put into stock waiting to be sold

- reduces the amount of money tied up in stock.

In your coursework project you will need to:

- produce and use a detailed **work schedule** that specifies the materials and components you will use

- use a time plan to set deadlines for the different stages of manufacture.

the **key people** involved in the production process include:

- clients (customers) who decide what needs to be made and raise the money for it
- marketing personnel who check out what will sell, to whom and at what price
- designers who agree a design brief with the client and produce ideas based on the product specification and available budget
- engineers who may need to convert a complex part of the design into a workable product at the lowest cost
- manufacturers who set up an effective production system to manufacture safe, high quality products at a profit
- retailers who distribute and sell the product
- consumers who buy and use the finished product and decide if it's value for money and does what they need.

labelling provides information about a product such as the brand name, product type, material, hazard warnings, country of origin and maintenance instructions. For example, a mains operated electrical product, such as a hedge cutter, will usually have details on its body about its electrical limits. It may also have information related to safety in use such as a small picture of goggles, an information symbol (read the book sign), a 'do not use in wet weather' sign and an electric shock hazard sign for frayed cable.

Some labels have a bar code that gives the manufacturer or retailer a product reference code, order number, client and manufacturing batch. The bar code can be used for valuable sales information that can be sent to the manufacturer for re-ordering, or for tracing faults.

In your coursework project you will need to:

■ use tools and equipment accurately, checking **dimensions** and **tolerances** (allowable variations in dimensions)

■ understand how computer software can be used for activities such as developing layout plans

■ understand that **systems** are used in industry to control design and manufacturing processes and **machines**.

a **layout plan** is important for working out the most economical way of cutting out parts from flat sheet material to reduce waste. Many parts for high volume production are now manufactured from sheet materials such as MDF and chipboard. These boards come in standard sizes so it is very important to lay the parts out economically on the sheet. Sometimes this can be very difficult if the sheet has a surface pattern or grain and the parts have to lay in a particular direction. If high volumes are being produced then waste can represent a considerable cost, including material and storage costs and the cost of waste disposal.

It is possible to estimate materials requirements by cutting out paper patterns of the parts, either full size or to scale, and arranging them on a table or the floor. It can be quicker to use computer graphics software and layout the parts to scale on screen.

There are a number of ways of cutting out parts from sheet material by computer numerically controlled (CNC) machines, depending on the complexity and the material:

- CNC routers and laser cutters can be used on wood and plastic sheet

Understanding Industrial Practices: Resistant Materials Technology © Nelson Thornes 2004

- mills and plasma arc cutting equipment can be used on metal
- sheet materials can also be cut by hand using a jig saw.

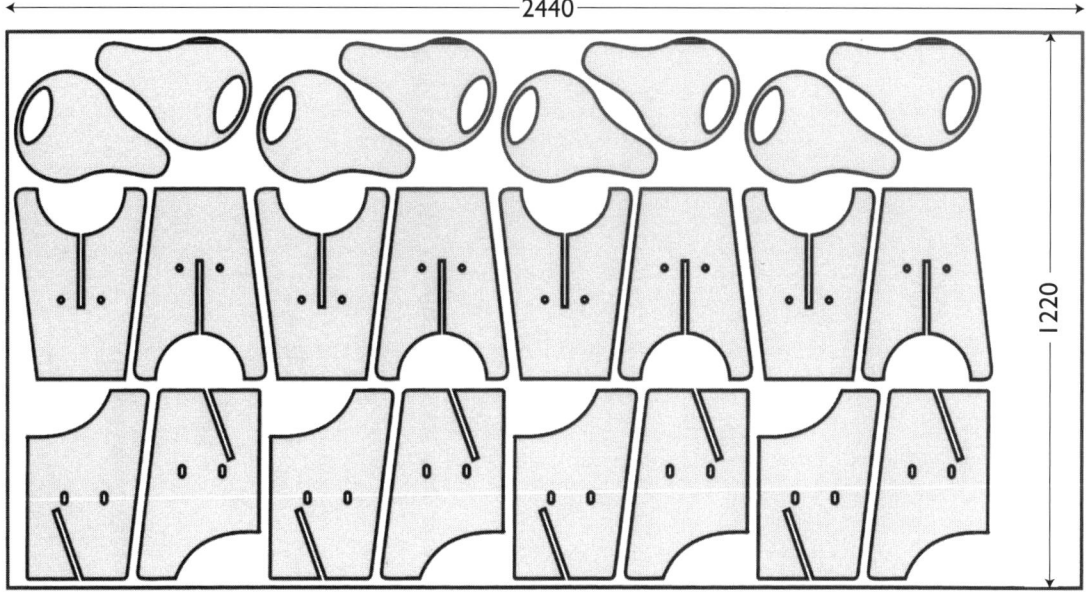

A layout for eight flat-pack seats on a standard size sheet of MDF

Coursework checkpoint: *layout plan*

Layout plans enable you to work out the amount and cost of material required for your product.

- ○ Estimate how much material you need by printing scale paper models of the parts and laying them out on a scaled standard sheet of material.
- ○ Try rearranging the parts for best economy.
- ○ Once you have the most economical plan, draw or take a digital photograph of this to keep as a record in your coursework folder.
- ○ Alternatively, use computer software to calculate and record your layout plan.
- ○ Calculate the cost of your material.

legislation for consumer products provides a set of minimum legal standards that **manufacturers** must follow to provide products that are safe to use.

Legislation such as the Health and Safety at Work Act 1974 protects employees in the workplace. You need to be aware of **safety** in the workplace and consumer safety when using products. For example, students designing children's toys need to follow the same requirements as industry and ensure that the products are made from non-toxic **materials** and **finishes** and have parts that cannot be swallowed. Consumer safety also relates to **fastenings**, which must not contain nickel or other toxic substances in case they are put in the mouth.

In your coursework project you will need to:

■ check that your product will be safe for your intended user.

▶

Coursework checkpoint: *legislation*

- ○ Collect examples of product labels or instructions for use from toys, electrical equipment (particularly for outdoor use) and household goods.
- ○ Find a label for a product similar to the one you intend to produce.
- ○ Use similar **labelling** or instructions for your own product.

life cycle assessment (LCA) means assessing the effect a product has on people or the environment from cradle to grave. It means investigating every aspect of the product's design, manufacture, use and disposal. LCA is used by the Ecolabelling Scheme to assess the impact that products have on the environment.

Coursework checkpoint: *life cycle assessment (LCA)*

Use the following questions to assess the impact your product has on people or the environment:

○ Are the raw materials renewable?

○ How much **energy** is used during manufacture?

○ Does the manufacturing process cause risk to people or the environment?

○ Does the distribution and packaging of the product cause risk to people or the environment?

○ Is the product safe to use?

○ Can the product be disposed of safely?

○ Does its disposal cause risk to people or the environment?

○ Can the product be recycled?

lifestyle marketing is where manufacturers and retailers target potential consumers and match their needs with consumer products. Market research is carried out to identify the buying behaviour, taste and lifestyle of the potential consumers. This establishes the amount of money they have to spend, their age group and which products they like to buy. New consumer products can then be developed to match consumer needs. A marketing plan is developed to promote the product so that consumers are aware of what is available.

the range of **machines** for production in resistant materials is very wide. They include:

● lathes (manual and **computer numerically controlled (CNC)**). Used to machine component parts using a rotating cutting action.

● mills (manual and CNC). Used to machine outline shapes or 3D surfaces.

● routers (manual and CNC). Used to machine profiles or 3D surfaces. Not suitable for metal.

● saws (circular, band, jig, donkey). Used for cutting. Blade type depends on material to be cut.

● buffers. Used for polishing materials.

● sanders/grinders. Used for removing small amounts of material, usually for shaping and finishing.

Understanding Industrial Practices: Resistant Materials Technology © Nelson Thornes 2004

- drills. Used for cutting holes.
- vacuum formers. Used with flat sheet plastics to produce 3D forms.
- injection moulders. Used with plastics to make complex 3D forms very quickly in large numbers.
- extruders. Used to produce continuous lengths of material such as pipes, mainly in plastics but also some metals.
- hand-held power tools. Can do many of the above **production processes** but require special **health and safety** precautions due to trailing flex and general lack of stability.

manufacturers need to make a profit, otherwise they will go out of business. They employ **key people** in the production team to design and manufacture the product. Their objective is to 'value-engineer' the product, which means parts are reduced to an absolute minimum to reduce costs as long as the product will still be reliable. They also put a planned maintenance process in place to make sure production can meet demand without coming to a halt.

The role of the manufacturer is to:

- be aware of **target market** needs
- devise a production plan
- set up a cost-effective **manufacturing system** that can meet demand
- reduce the number of component parts and simplify **assembly**
- reduce labour and material costs
- produce consistent products using a **quality assurance (QA)** system
- manufacture products at a profit
- manufacture products that are safe for employees, **consumers** and the environment.

a manufacturing specification ensures that a product is manufactured as the designer intends. It provides clear and detailed instructions about the product's styling, materials, construction, dimensions and tolerance limits. The manufacturing specification is an essential part of the production plan and enables the profitable manufacture of identical products. It is used as a standard document for checking the quality of each product.

A manufacturing specification should include:

- a written description of the product, including performance criteria
- information about bought-in component parts, fastenings or fixtures
- **working drawings** to show dimensioned views of all the component parts, including critical **tolerance** limits
- a materials and component **parts list**
- details of **colour** and applied **finish** or surface texture.

This information enables a sample product to be made. Manufacturing specification documents are often produced and stored on a computer database, making it easier and quicker for all the production team to access information about the product.

In your coursework project you will need to:
■ develop and use detailed specifications
■ use a manufacturing specification to evaluate the quality of your product.

a **manufacturing system** is decided by the type and number of products to be made. The manufacturing system is chosen to enable **manufacturers** to produce products quickly and efficiently. Different systems include:

- one-off, jobbing production (low volume such as 1 to 20)
- batch production (middle to **high volume** such as 20 to 5,000)
- high volume, **mass production, flow-line production** (high to very high volume such as 5,000 to 100,000)
- **continuous production** (very high volume such as 100,000 or above).

The **method of production** used depends on the product size, complexity and the production process involved. For example:

- plastic drain pipes are made by continuous production because extruding machines need to heat up to exact temperatures. They take a long time to shut down and start up again.
- small washers for nuts and bolts can be made in batches up to 100,000 because they are a single part. Stamping machines can be quickly changed to stamp a different item.

With careful design and the use of appropriate equipment, most products could be batch produced in school workshops.

market research is an industrial practice, carried out by **manufacturers** to identify the buying behaviour, taste and lifestyle of potential **consumers**. A market research report can establish the size of the **target market group**, the product to be developed, product price ranges and the competition from other manufacturers.

1 Primary market research can be collected through:

- a trip report to identify product information and new ideas
- **product analysis** of similar products to find out about **materials** and components, **quality**, **safety** and value for money
- a user trip to **test** could be improved
- questionnaires asking intended users about the products they buy and their likes and dislikes
- industrial visits, exhibitions
- interviews with experts such as **designers**, manufacturers, teachers, other adults.

A trip report can provide useful information about the product to be developed, the intended target market group, **style** trends, new ideas, design details and the product price range. A trip report can involve:

- window shopping
- going into stores to look for ideas, trends, themes and styles
- going to art galleries, clubs, craft fairs, and museums to look for ideas about art, sculpture and music.

Buying behaviour can establish the profile of a potential target market group, such as:

- the amount of money they have to spend
- the age group, their taste and personality influenced by nationality, lifestyle, family group and **brand** loyalty
- the products they like to buy, e.g. domestic products, luxury goods.

In your coursework project you will need to:

- take account of a range of users

- use artistic, cultural, social and environmental influences when developing your design ideas

- use the influence of past cultures and styles and the work of other designers when developing ideas

- be flexible in the way that you respond to new ideas and new opportunities.

---

2 Secondary market research can be collected from existing information in:

- magazines, newspapers, textbooks
- TV programmes
- CD ROMs and databases, the Internet.

Coursework checkpoint: *market research*

Use market research techniques to find out style trends for your intended product. You can do this by producing a trip report.

1 Decide which stores you will visit depending on your product type, e.g. department or electrical store, sports or toy shop, market.

2 Decide what information you need to find out.

3 Ask permission from the manager, if necessary, explaining why you're doing a report. Ask permission to take photos.

4 Make sketches and notes on products about:

- materials, **fixings**, **fastenings** and components.
- **colour** and style trends
- fixings, fastenings and components
- production methods
- **product maintenance**
- price ranges.

5 Compare your trip report with that of another student who aims to design a similar product to you.

6 Use the information you have collected to give you ideas about design and style trends.

7 Try to spot new trends that you could incorporate into your own designs.

8 Use your trip report information to develop a **mood board** for your product.

---

marketing is a research and sales **system** that makes sure that there is a market for a product before it is made. Marketing information is later used to promote sales of the manufactured product. It involves developing a sales plan for a product aimed at the **target market group**. Marketing is also about developing a competitive edge by providing reliable, high **quality** products at a price **consumers** can afford, combined with the image they want the product to give them. This is sometimes called **lifestyle marketing**.

1 A successful marketing plan uses **market research** to find out about:

- user needs
- consumer demand
- the age, income and location of the market
- the size of the market
- the product type consumers want
- the price range consumers will pay
- trends affecting the market
- competitors' products and marketing style
- the timescale available for selling into the market.

In your coursework project you will need to:

- take account of a range of consumers and their needs and wants

- use cultural and social influences when developing your design ideas

- use the influence of past cultures and styles and the work of other **designers** when developing ideas

- be aware of the environmental impact of marketing new products

- be flexible in the way that you respond to new ideas and new opportunities.

▶

2 A marketing plan can involve the advertising and promotion of **brands** through **retailers**, newspapers, magazines, posters, hoardings, TV, radio and the Internet.

Some manufacturers use different brand names for similar products to give them a more exclusive image for a different target market group. For example:

- Panasonic manufacture a range of audio systems including the more expensive 'Technics' systems.
- Most sunglasses have the same technical specifications but are priced according to brand name and style.

Many products use standard parts but are repackaged to maintain sales. For example, marketing departments use trend and style information to develop new products in order to maintain product sales and profits. The 'throw away society' is not always good for resources and the environment but it does keep people in work.

Coursework checkpoint: *marketing*

Collect product advertisements from magazines and try to identify the target market group the product is aimed at.

- ○ Is a particular product type aimed at a specific age range?
- ○ Is the product advertised using different media such as TV or street hoardings?
- ○ Does the advertisement try to promote an image about the product?
- ○ Is the product aimed at a particular lifestyle?
- ○ Is this product lifestyle or image to be found in the shops that sell these products?
- ○ Where does your product fit in with image and lifestyle marketing?

---

mass production (high volume production) is used for manufacturing large quantities of products for stock or order.

materials and components can be chosen from a wide range to match the needs of products and users. Almost any resistant material can be used for your own project work as long as it is available and its **properties** match the product specification. New resistant materials are continually being developed so you should keep up to date by looking in newspapers, magazines and on the Internet. Although you need to be aware of these new materials you are not required to use them in your own work. You can generally find what you require in the school workshop store. However these materials may not be in the correct sizes or quantities so check this out before deciding on construction.

1 Wood is very eco-friendly as a material as trees can be replanted over a period of time and take little energy to harvest. Trees also provide the world with some of the oxygen we breathe.

- Natural woods fall into two groups, hardwood and softwood. Natural solid wood is warm in appearance and is strongest along its direction of grain.
- Manufactured wooden boards include MDF, chipboard and ply. MDF is used by nearly every flat-pack cabinet manufacturer and is excellent for computer numerically controlled (CNC) batch produced products.

28

29

In your coursework project you will need to understand:

- the properties and characteristics of a range of specific materials
- that combining materials can give improved properties
- that resistant materials can be cut, shaped, combined and processed to create more useful properties
- that the properties of resistant materials should be suitable for the **end-use**.

2 All metals are mined from the earth so take a lot of energy to produce, making them not as eco-friendly as wood.

- Metals such as steel, aluminium, copper and brass are commonly found in most workshops. They are used extensively by industry in sheet, rod, strip, moulded or cast forms.

3 Plastics are made from oil which is non-renewable so plastic is not an eco-friendly material. However, we would now find it difficult to survive without plastic products.

- Common plastics include acrylic, polythene, polypropylene, PVC, polystyrene, ABS and nylon.
- Most thermo-plastics can be injection moulded or vacuum formed into complex 3D shapes. Thermo-setting plastics, such as polyester resins, cannot be remoulded after they set. Each different plastic has its own special characteristics so needs to be chosen carefully when designing plastics products.

Coursework checkpoint: *materials*

Materials can be used in their original state or can be combined, joined or layered to provide improved properties. Experiment with different techniques to improve the performance or appearance of resistant materials. You can try:

- using joining processes to change the properties of materials
- using triangulation to improve the strength of constructions
- adding high strength components such as corner brackets or supports to otherwise lightweight but flexible constructions such as shelves
- adding components for decorative effects (eyes on toys or vinyl letters on products)
- laminating materials to retain **grain** direction and improve strength
- moulding, casting or shaping materials to change their shape without **joining**
- using surface coatings for better handling, stain, water or wind resistance.

---

 a **method of production** is decided by the type and number of products to be made and whether they are made to order for immediate delivery or for stock.

 The increasing use of Information and Communication Technology (ICT) has enabled the development of new methods of production, such as quick response manufacturing (QRM). This involves manufacturing teams making products to order, using batch production.

Methods of production can include:

- **one-off** (low volume or jobbing production)
- batch production for stock or order, often using quick response methods
- high volume (mass production, flow-line production) for stock or order
- continuous production (continuous flow) this process is used for high volume production.

In your coursework project you will need to:

- design a product that could be manufactured in quantity
- choose the most suitable method of production for your product and give reasons for choosing it
- understand what is meant by one-off, batch production, high volume and continuous production
- be able to give examples of products produced by each method of production
- explain the advantages and disadvantages of each method of production
- understand how **quality assurance (QA)** systems and **quality control (QC)** techniques are used to manufacture high **quality** products.

micro-processors are the brain of modern computers and control the system's operation, storing and analysing data. They work with other chips to provide input, output and memory functions. Computers have increased the speed of change and the way we live by processing and communicating data on a global basis and by automating many machine processes. Most, if not all, manufacturing industries now use computers to store, sort and communicate data to help them remain competitive and increase their product market share.

Programmable Logic Control (PLC) chips are complete computers in a chip. They have a built in memory and a number of input and output terminals. These can take information from sensors and drive output stages to control devices such as motors, lights, solenoid valves, relays and warning devices.

- PLCs are easy to use and can be programmed in the same way as a computer.
- They are low cost when compared to mechanical systems.
- Once set up they are extremely reliable as there are no moving parts.

PLCs have replaced mechanical control and timing systems in many machines, both in industry and in homes. For example they are used to control:

- automated industrial systems such as robotic handling and computer controlled machines
- domestic microwaves, washing machines and heating systems
- smaller consumer goods such as battery chargers and mechanical toys.

modelling in 2D and 3D is a key industrial practice because it enables manufacturers to test and modify a prototype product before putting it into production. This saves time and reduces costs. For example, a large construction such as a garden seat can easily be modelled in a smaller scale to check it for visual appearance. The cups shown below right were modelled full size in painted wood to test for colour, shape and feel.

32

Product modelling

1. Designers use 2D and 3D modelling to visualise ideas, either by drawing or sketching by hand, or by using **computer-aided design (CAD)** software. CAD software can also be used to model and colour products on screen.

2. Layouts for cutting sheet material can also be tested by modelling in paper or on screen by computer, either full size for small products or to a smaller scale for large ones. Pieces are placed on the material in the most economical way to reduce waste and to cost the amount of material required.

3. Modelling can involve trialling **joints**, **fastenings** or new techniques. It is important to make use of prior knowledge in your coursework and to only trial processes that you are intending to use for your project. Put a note in your coursework folder to explain your skill and practise any new technique you need to use to manufacture your product.

4. If the final product has a mechanism or electronic system inside it, this too is usually modelled to see if it works. Case details are then finalised. The case is usually part of the structure to hold the mechanism, as this is cheaper to produce than having a frame and another case around it. Kettles are a very good example of this.

5. Many designers model small products in 3D to visualise and handle the product before production takes place. The fastest method is used to make the model in foam, MDF or other 'soft' or easily worked materials. Models of this type are finished to make them look realistic, even though they may not work.

modules/modular construction are often used in industry to make complex products. This means that a section of the product can be easily changed or extended by using add-on parts. For example, many expensive cameras have interchangeable lenses and bodies so could be said to be of modular construction.

Modular also means that a product construction is made from the same type of part that is repeated many times. Kitchen cabinets are made using this principle. The cabinets are all the same height and depth but come in a range of widths to enable a combination of them to fit all kitchens. This modular system keeps cost down, speeds up production and makes the cabinets easier to fit or replace if one is damaged.

You may be able to use a modular system to speed up the production of your project by having several of its parts made in an identical way using either **computer numerically controlled (CNC) machines** or **jigs**.

mood boards are used by many professional designers to explore colour and **style** ideas for manufactured products. They are used to generate first ideas, illustrate and explain themes, product styling and colour ranges. Mood boards provide useful starting points for the design of children's toys and domestic products including furniture and matching product ranges. Some designers use the same sort of rapid layout technique for collecting ideas for mechanisms, **fastenings** and fixtures. These may also influence the final style of the product as well as its operational qualities.

In your coursework project you will need to:

- use hand techniques or computer software for drawing, designing, colouring and modelling ideas
- use hand techniques or computer software for developing layout plans and costs
- work out the degree of accuracy needed for your product to fit well together.

In your coursework project you will need to be aware that:

- making parts the same size and shape can speed up production
- using CNC machines and/or jigs can increase accuracy and speed up production.

In your coursework project you will need to:

- take account of a range of users
- use cultural and social influences when developing your design ideas
- consider the influence of traditional and modern designs when developing ideas

▶

Mood boards are quick to produce and can be made from many types of image such as:

● magazine 'swipes', photos, colours from paint charts, tissue paper

● architecture, art or style images

● anything else that provides inspiration for the product.

A mood board provides a picture backdrop that is often put into a colour or style group, so there may be a neutral, pastel, bright or dark, soft or sharp colour story. Often it is the shape or style of the displayed items that provides inspiration. The purpose of mood boards is to provide starting points for generating product ideas so they must be done very rapidly.

Coursework checkpoint: *mood board*

Remember that a mood board is *not* a scrapbook! Be very selective in choosing what to include by finding items that best fit the mood or style you are trying to achieve. Remember to *annotate* your mood board with headings or brief sentences to explain the story or mood surrounding your product.

market opportunities generally fall into the categories of **needs and wants**. For example, you may need clothes to keep you warm but you may want them to be fashionable. You may need to tell the time and have a clock or watch but many people want to own more than one of these because styles change. Commercial **manufacturers** have developed very sophisticated ways of selling products to customers and most of what is offered is by market push rather than market pull.

● Market push is when people are encouraged by manufacturers through advertising to want products they don't really need but which could be fashionable, a novelty or very pleasurable to own.

● Market pull describes when people actually need a certain type of product and actively go about searching for it. Manufacturers try to pick up on this and produce the products potential consumers want at a price they are willing to pay.

Coursework checkpoint: *needs and wants*

○ Test out your ideas for your product on potential consumers before production.

○ Develop a **marketing** strategy for your product.

a **net** is another name for a surface development. For example, the net of a cylindrical tube would be a rectangle. Sometimes 3D models can be constructed in card from a net of the product. It is only really feasible to construct a product from its net if it has flat sides. Nets are used extensively in the design of product packages for distribution and sales. Carefully designed nets are used to minimise the amount of material used, yet offer a secure, attractive container to protect the product from damage in transit.

one-off production (custom made, jobbing production) means designing and making a single product to a **client's** requirements. Sometimes a customer might require a slight product variation and one-off production allows for this. For example, a chair might have a lower or higher back depending on customer preference, yet still be in exactly the same style.

Designing and making one-off products is usually very labour intensive, using skilled workers where **quality** is checked as the work progresses. One-off products are usually more expensive than **batch** or mass-produced items because of the amount of time spent designing and making them. One-off resistant materials products can include:

- trophies made for special presentations
- individual constructions for exhibition, foyer and shop displays
- hand-made toys such as rocking horses
- individual furniture and high quality games.

More designers are beginning to use **computer-aided design (CAD)** and **computer-aided manufacture (CAM)** to speed up product development and increase product quality for one-off production. Modern computer technology and **quick response manufacturing (QRM)** can also make it possible for **high volume** product manufacturers to make one-off products to order. These could be less expensive than buying traditional one-off products. For example, it is possible for a furniture manufacturer to:

- take a customer's requirements for a new design based on a standard product
- customise the dimensions using a basic CAD drawing of the product
- make the product as a one-off using CAM
- deliver the product to the customer within a couple of weeks
- code the product ready for re-ordering.

Your coursework product may be made as a one-off, but you will need to suggest how they could also be produced in quantity.

a **parts list** or materials list is a table used to record the part, size, material, colour and finish required to make a complete product or sub-assembly. It is usually part of a **working drawing**, which itself is part of the **manufacturing specification**. A colour column is normally only needed for plastic or painted parts. Some parts lists may include costs and locations for bought-in component parts, such as the clock mechanism in the parts list, shown on page 48:

▶

In your coursework project you will need to:

- choose the most suitable **method of production** for your product and give reasons for choosing it

- understand what is meant by one-off production, why it is used for manufacturing products and how it compares with other methods of production

- understand that using **standard components** can make manufacturing more cost-effective

- understand how **quality control** is used to manufacture a high quality product.

Part	No.	Length	Width	Thick	Material	Colour	Finish
Face	1	125	125	10	Walnut	Natural	Wax
Outer upright	2	155	20	15	Sycamore	Natural	Wax
Centre bottom	1	72	60	18	Mahogany	Natural	Wax
Centre top	1	23	60	3	Mahogany	Natural	Wax
Feet	2	75	15–0.1	dia.	Walnut	Natural	Wax
Clock hands	2	60/45	8	0.2	Aluminium	Black	Painted
Clock mechanism	1	60	60	12			

Parts list for small wooden clock (see worksheet 4 for a drawing of this clock)

The **tolerance** limits for all the component parts are not easy to show in a parts list, as each part may need a tolerance for each dimension (L x W x T). Critical tolerances are normally shown on the working drawing. They usually relate to component parts that have to fit together, such as the feet for the clock in the above table.

percentiles are values representing the percentage of the population at or below certain measurements.

For a design to be successful it must suit the sizes and proportions of the people who use it. A product designed for the average sized man may not work well for small females or children. However, it is impractical to design all products to suit all people, so designers use **anthropometric** data to decide what percentage of the population a product will work for. For example, ride on toys for children between the ages of 2 and 5 are designed to suit about 10 % of the population (3 to 13 percentile height values). The 3 percentile value is the height of the smallest child who could just manage to get his or her feet to the floor. The 13 percentile value is the height of the tallest child who could comfortably use the product. Below 3 % of the population would be too small and above 13 % would be too tall.

> **Coursework checkpoint:** *percentile*
>
> Take and record body measurements for a sample of people in your **target market group**.

performance requirements are functional specifications such as 'strong' or 'lightweight' that relate to a product's function and **end-use**. For example, the performance requirements of a biker helmet may be crash resistant, lightweight, water/windproof.

In industry performance requirements are part of both the **design and manufacturing specifications**. They are normally strictly observed by **testing prototypes** before the final product is released to the market. It is very important that **health and safety legislation** is taken into account when designing all products, especially where **safety** is concerned.

Understanding Industrial Practices: Resistant Materials Technology © Nelson Thornes 2004

product analysis is an important industrial process that provides useful information about the design and manufacture of existing commercial products. The activity includes the practical analysis of products as well as the collection of data from catalogues, visits, experts, Internet, etc.

Designers undertake product analysis to help them evaluate:

- the price ranges of products
- the processes used to manufacture the product
- the **properties** of **materials** and components
- the **quality** of design and manufacture
- the **fitness for purpose** of the product for the **target market group**
- why the product is successful.

Designers also use product analysis to develop ideas for new products and produce new product **design specifications**. Product analysis will help with the design of your own coursework product.

One way of evaluating a product is to use similar criteria that you use when developing a product design specification:

- Explain the product's **end-use**, its function or purpose.
- State the target market group and describe their needs and values.
- Describe what the product looks like – its **aesthetic** characteristics, the image it gives the user.
- Describe how the product performs – the **functional** characteristics of the product, materials and components, including maintenance.
- Examine the manufactured **quality** of the product – such as materials, **construction** and **fastenings**.
- Investigate how safe the product is to use or for people and the environment.
- Finally, explain why the product is suitable for the target market group. Include references to any special features, how well it is designed and made and the value for money of the product.

Coursework checkpoint: *product analysis*

Analyse **one-off**, **batch produced** and **high volume** products, to find out why and how they are made by these different production methods.

When you analyse a product remember to:

- ○ sketch the product
- ○ identify the materials and processes used to make the product
- ○ investigate the properties of the materials used and **product maintenance** requirements
- ○ work out and sketch the component parts required to make the product
- ○ work out the order of **assembly** of the product.

In your coursework project you will need to:

- take account of the needs and values of a range of users
- understand that changing fashions or **style**, the price consumers will pay, **brand** image, manufacturing techniques and **environmental issues** can influence the design and manufacture of products
- understand the difference between **quality of design** and **quality of manufacture**.

product data management (PDM) software organises and communicates accurate, up-to-date information in a database, monitors production and supports fast, efficient and cost-effective product manufacturing on a global scale. This is made possible by exchanging data through a network of computers in different locations.

PDM is used by product **manufacturers** as part of a **concurrent design and manufacturing (CDM)** system, to handle and make effective use of data from a range of **computer systems**, including **computer-aided design (CAD)**, computer integrated manufacture (CIM) systems and **electronic point of sale (EPOS)** sales systems.

PDM systems use software that enables research data, **style** and **colour** information, design ideas and images to be input into the computer. Data can be input from digital and video cameras, photo-CDs, scanners, 35mm slides, spreadsheets and CAD software. **Virtual products** can be developed on screen and sent to **clients** for approval. The image of the product and all the technical data needed for its manufacture can be exchanged or transferred to different stations using PDM systems.

PDM software enables access to product information for every member of the **product development team**. When changes are made to design, **materials** or **costing**, the information can be automatically sent to each team member. This means that everyone, including the **designer**, manufacturer, costing department and materials planner is working with the latest information.

PDM reduces the product time to market and enables manufacturers to use **just in time (JIT) methods of production** and **quick response manufacturing (QRM)**.

PDM software enables communications using **Information and Communication Technology (ICT)** between departments, manufacturers and **retailers** in the same or different city or country. This enables the design of products in one location and manufacturing in another, resulting in a **global manufacturing** system. For example, designers can research product information in one country and download from a laptop computer via electronic links to the central design office thousands of miles away. Product development can begin immediately. The final product specification can be sent electronically to the manufacturing site in another country. **Quality** can be monitored on the production line using digital cameras and sensors linked to the central design office.

product development takes into consideration current trends and the potential market for the product. Many new products are developed from older, well-proven component parts or products, which are remodelled to meet changing **consumer** taste. The market for music systems such as CD players is like this, where the product mechanics may have a working lifespan of more than five years but styling features are changed much more often. New technologies have an effect on product **development**, especially if the product has an electronic system in it. Computer chips are now used to control most automatic, or semi-automatic products such as washing machines, microwave cookers, mobile phones and even battery chargers.

The introduction of new **materials** changes the way products are developed and how they look. For example, products that were originally made in metal and later developed in plastic look quite different because of the types of processes used. An example of this is the domestic kettle, many of which now have plastic shells.

Important aspects of product development include:

- new technologies (includes materials, fixings, processes, control systems (mechanical and electrical), structures, mechanisms)
- market trends (forecasting consumer needs or buying trends)
- modelling (testing ideas before production. These ideas could be virtual or real.)
- value engineering (reducing the number of parts and cost without sacrificing reliability).

product development teams include those who work in market research, design, accounts, production, quality systems, sales, marketing and distribution.

product maintenance needs to be taken into account when designing and manufacturing products. Consumers want high quality, well-designed products that require minimum maintenance. Product leaflets and labels provide consumers with instructions about use, safety and maintenance.

For example, it should be easy to change a dust filter in a vacuum cleaner or the battery in portable electrical equipment such as power tools, torches or clocks. Some machines need to be oiled at regular intervals. All cars are required by law to have routine maintenance safety checks carried out every year after the car is three years old. Most products have instructions on how to keep the product clean.

1 Product maintenance is related to the life expectancy of the product and its end-use. For example:

- The seat belt mechanisms of cars are expected to last the lifetime of the car and require minimum product maintenance.
- Most clocks and watches with springs need winding up each day, whereas those with batteries only need maintenance when the hour changes or about every year to change the battery.
- Most products with mechanisms come with advice on how to keep the product in good condition, even if it only means keeping it washed or clean (tin-openers, power tools, dish-washers, scissors, etc).

2 New materials and systems are continually being developed, to provide products that require minimum maintenance. For example:

- non-stick PTFE plastic is used for coating saucepans
- low-friction, maintenance-free plastic bearings are used in personal hi-fis and other equipment with low speed shafts
- micro-processor and programmable logic control (PLC) chips are used to replace clockwork or mechanical control and timing mechanisms. Electronic control systems rarely have moving parts so they are extremely reliable and almost maintenance free.
- LED solid state lights and fluorescent lamps are beginning to replace light bulbs in a wide variety of equipment because they last much longer.

In your coursework project you will need to:

- think about maintenance when designing your product, making sure that it will be easy to look after by your **target market group**
- make sure that you choose materials that will be easy for the user to clean. For example, choose materials that are easy to wipe clean and have a non-toxic **finish** for a child's toy.
- make sure that your product has easy to use functions. For example, make it easy to change the battery in a clock or electrical toy.

a **production line** or assembly line is where high volume products are manufactured by teams of people in a factory. Each production process follows on from another. A production line is part of a **production system** to manufacture products quickly and efficiently.

In industry a production line requires one operator to perform one task continuously, such as assembling the final product. Other people in the production line perform different tasks repeatedly. The work flows in a line through a series of workstations, where each process takes a similar amount of time. Production lines are used less frequently in the new industries because this **method of production** can be very tedious and repetitive. It can lead to lower levels of motivation and productivity because machine operators do not feel like important team members. Production line working is being replaced by **cellular manufacture** where each team member produces a complete product.

production line simulation involves working on an exercise as part of a production team to manufacture a product in high volume. This could be something relatively simple such as making a bird-box where each team member has just one task to complete. Production simulation can also be achieved using computer software to develop and model products.

production planning means considering the ways a product can be manufactured at the lowest cost without reducing **quality**. A production plan is used to choose the best, safest and most cost-effective processes, the best layout for equipment, **materials** and people, and the best ways to control a product's quality.

A production plan provides clear and detailed instructions for the manufacture of the product. It is part of a **quality system** used by **manufacturers**. The production plan consists of standard documents that show all the information needed to make the products. Different products and different **scales of production** need different types of production plans, so manufacturers develop standard **templates** to meet their own requirements.

A production plan for a plastics waste-pipe manufacture would include the following information:

● The **manufacturing specification** that includes a **working drawing** with details of materials, components, dimensions and **tolerance** limits.

● The **work schedule** with details of manufacturing processes and the sequence of **assembly** (for example, putting the rubber ring seals in the couplings). This work schedule would also include the technical requirements such as machine timings and heating cycles.

● A **flow diagram** that shows where and how to check for product quality.

The production plan forms a key part of the **quality assurance** (QA) system, because it documents each stage of manufacture. This enables each product to be made to the same quality.

Understanding Industrial Practices: Resistant Materials Technology © Nelson Thornes 2004

- Checks are made at each stage of production to monitor quality.
- If there are problems during production, or faults occur, they will be identified quickly. This **system** provides **feedback** so that if necessary, changes can be made to the production plan or to the design of the product.

A production plan can be complicated because production teams require a lot of information. Many manufacturers use **product data management (PDM)** software to monitor production. This means that if any changes are made in the production plan, by any of the production team, the information is immediately available to everyone in the team.

Coursework checkpoint: *production plan*

I Draw up a production plan for your coursework product. Include the following information:
   ○ A manufacturing specification that provides clear and detailed instructions about the product's design, materials, **construction**, components, dimensions and **tolerance** limits.
   ○ A work schedule with details of manufacturing processes and the order of assembly.
   ○ Any technical requirements for setting up the machines (such as **computer numerically controlled (CNC)** routers or lathes).
   ○ A flow diagram to show where and how you will check your product quality.
2 Use the information in your production plan to monitor the quality of your product, making checks at each stage of production.

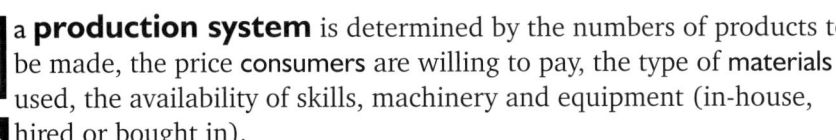

a **production system** is determined by the numbers of products to be made, the price **consumers** are willing to pay, the type of **materials** used, the availability of skills, machinery and equipment (in-house, hired or bought in).

Production systems enable **manufacturers** to produce goods quickly and efficiently and can include:

- one-off production (low volume or jobbing production)
- batch production
- high volume (mass production, flow-line production)
- continuous production.

many **production techniques and processes** are used with resistant materials. Most are dependant on the equipment or **machines** available. **Health and safety** is important when using any machine or hazardous process.

The main techniques include general cutting or wasting, **joining** and holding plus:

- wood: fabrication, moulding/pressing, laminating
- metal: fabrication (joining separate parts), pressing, forging, extrusion, spinning, casting
- plastics: fabrication, injection moulding, compression moulding, extrusion, vacuum forming, casting.

The machines used for processing resistant materials include:

- Wood: drills; circular saws; jig saws; band-saws; routers; belt-sanders; lathe; planers; planer-thicknessers; and use of **computer** ▶

In your coursework project you will need to:

■ understand a range of manufacturing processes and techniques

■ understand which manufacturing processes and techniques are suited to which materials and how these can be applied

■ understand how to manufacture products in quantity using the appropriate processes and techniques.

numerically controlled (CNC) equipment including router/laser cutting.

- Metal: drills; circular saws; jig saws; band-saws; belt-sanders; lathe; mills; welders; extruders; presses; casting; forges and basic use of CNC equipment (lathe, mill).

- Plastics: drills; circular saws; jig-saws; band-saws; routers; belt-sanders; lathe; laser cutters; strip heaters; buffers; injection moulders; blow moulders; vacuum formers; extruders and vinyl cutters (letters, logos, etc).

Programmable Logic Control (PLC) chips are complete computers in a chip. (See micro-processors.)

the **properties** of a resistant material depend on the basic raw materials from which it's made, the finishing processes used and the constructions used to join them together. Products can be made from basic raw materials such as hardwood or steel, from composite like plywood or carbon fibre, or from metal alloys like brass or solder. Combining materials improves their properties, so composites are usually stronger than their individual component raw materials.

New materials are continually being developed for new end-uses. For example, PTFE was developed as a high specification, high cost product for use in space travel. PTFE is now available at a reasonable cost for use on non-stick pans. It can withstand high temperatures and has low frictional properties.

When designing your own products you need to know exactly what properties the product needs so that you can choose appropriate materials to make it from.

Properties can be:

- **aesthetic**, such as soft, hard, shiny or dull, relating to a material's appearance or feel

- **functional**, such as water resistant, tough, durable, slippery or flexible, relating to performance requirements.

It is sometimes useful to use more than one type of material when making a product so that different parts have different characteristics. Wood, metal and plastics can be combined to make products look or function better.

a **prototype** is a detailed 3D model or construction, usually made from inexpensive materials, and is used to test a product before putting it into production. Prototyping is a key part of the manufacturing process. It is used to trial a design, to see how materials, mechanisms or systems behave, to test for appearance, to try out assembly processes and to work out costs. In industry there are different ways of prototyping, depending on the level of production.

1 High volume manufacturing.

Non-working prototypes are modelled:

- as drawings either on paper as rendered (fully coloured and shaded) images

- on screen as virtual reality models

- as a real casing to test for feel, size and weight.

These methods are good for testing initial reaction to style and colour.

37

38

In your coursework project you will need to:

- understand the relationship between the properties of materials and manufacturing processes, so that you make the best use of materials

- understand that materials can be combined to create more useful properties

- make sure that the properties of your materials are suitable for the **end-use**.

In your coursework project you will need to:

- take account of **critical dimensions** when planning your product manufacture

- test and modify your product at critical points in its manufacture, and work out ways to improve its **quality**.

Working prototypes are modelled:

- full size to test for size, strength and function
- to test mechanisms and electronic systems for their operational qualities and reliability. Quite often this type of operational prototype does not need to look good to check its performance.

2 In more exclusive **one-off production**, products are made for individual **clients**.

- The prototyping process can be more complex because the design is made to the client's requirements and may be more unusual. The product is developed with client approval from the original ideas, often from paper or virtual reality screen-based images. A fully working prototype is sometimes the end product.
- In many cases the prototype is modified or developed during construction to make sure it complies with the agreed client **specifications**. The film industry uses many prototypes of this nature for new products that are viewed on screen. 'Star Wars' is a typical example of a film that uses prototype models for many of the sets including weapons, body armour and modelled spacecraft.

 producing something to a high **quality** means meeting set standards to make identical products that have a zero fault rate. When designing a high quality product a **designer** must produce a desirable product for the **target market** at the right price. The designer needs to consider **fitness for purpose**, value for money, the type of material available, the method of manufacture and the **style** of the product.

1 Quality for the consumer means having a product that is fit for purpose. This can be evaluated through its performance, price and **aesthetic** appeal.

2 Quality for a **manufacturer** means meeting the product **manufacturing specification** and finding a balance between the following:

- profitable manufacture of identical products on time and to budget
- the needs and expectations of the **consumer** and the environment.

3 Tight **tolerance** limits on component parts and high **performance requirements** result in a higher price and a higher level of quality. This requires high levels of training and skills and the prevention of faults through the use of **quality assurance** (QA) systems.

4 Wide tolerance limits on component parts and low performance requirements result in lower cost products and lower quality. Low quality manufacturing generally results in high levels of reworking and repairs.

In your coursework project you will need to:

- understand that the quality of a product can be judged by how well it meets user needs, the appropriate use of **materials** and components, how the product meets manufacturing and maintenance requirements, its fitness for purpose, its impact on the environment and its value for money

- evaluate the quality of your product against the **specification** to make sure that it is suitable for your intended market

- modify your product where appropriate to improve its quality so it meets the needs of your target market group.

quality assurance (QA) is a planned system that sets up quality procedures to make sure that products fulfil the required specifications. The aim of QA is to make identical products with zero faults. The QA system identifies where problems are likely to occur and sets up testing and quality control procedures to stop them happening.

QA systems are planned to monitor every stage of design and manufacture. They ensure that the product meets standards such as those set by the British Standards Institution (BSI). The BSI ensures that manufacturers make products that fulfil the safety and quality needs of consumers, and the environment. The labelling and advertising of products is monitored by UK and European consumer legislation.

To prevent faults, QA makes use of written procedures and systems such as:

- production plans
- specifications
- costings
- work schedules
- computer systems
- **quality control (QC)** systems
- standard working processes and practices
- **inspection** and fault finding.

QA systems require teams of people to work together and share information about the product. This often results in vast amounts of documentation and paperwork, so many product manufacturers use product data management (PDM) software. This monitors quality and production. PDM also enables the whole product team to have up-to-date information about the product.

quality control (QC) is used to test and monitor the manufacture of products.

QC involves checking tolerances at critical control points for accuracy and conformance to the specification. This ensures that the product meets consumer expectations. QC test methods, checks and inspection processes enable the manufacture of products that are identical, with a zero fault rate.

QC is the practical way of achieving quality assurance (QA). QC is the responsibility of everyone in the production team. It:

- makes use of specifications and standards (BS) to monitor and control quality
- makes use of inspection to identify faults
- is applied to raw materials, design, production and the finished product
- provides feedback to the quality assurance system to make sure that it is working properly.

1 Incoming raw materials and components are checked and tested for faults, correct size, colour, characteristics and performance.

2 The design and production planning departments develop a production plan to provide:

- clear specifications for materials, colours, component details, sizes and tolerances

- equipment that works well enough to meet the required tolerances
- performance checks to ensure tolerances are met using **quality indicators**
- inspection at regular stages
- clear manufacturing and **assembly** details
- quality checks for product assembly, packaging and **labelling**
- organised distribution in the correct quantities to the correct addresses.

3 **Production planning** applies quality control procedures which check against the design and manufacturing specifications, tolerance limits, **working drawings** and the **work schedule**. These procedures enable quality to be checked at **critical control points** in manufacture, in order to produce a perfect product. Quality inspectors use quality indicators that identify how the quality is checked. The quality system is used for identifying faults that are recorded and acted upon.

4 The first samples are checked against the **manufacturing specification** and any faults in manufacturing are put right.

5 The production run is then started and quality is checked at regular intervals.

6 The final inspection compares the product with a high quality perfect sample.

Coursework checkpoint: *quality control*

Remember to record any changes you make to the design, manufacturing processes, assembly or **finish** of your product. This will enable you to explain how you could manufacture perfect products in quantity.

---

43 **quality indicators** are used at **critical control points** in a product's manufacture to check how the **quality** of a product conforms to the manufacturing specification. Quality indicators are either variables or attributes.

Variables include anything that can be inspected by measurement such as length, height, width, diameter, angle, mass, pressure or temperature. In product manufacture, allowance is generally made for some variation in measurements because greater accuracy is more expensive to manufacture. Dimensions are 'variable' quality indicators that allow for variations in measurement within defined limits. This allowable variation in dimensions is called **tolerance**. For example, a component dimension with a tolerance of 100 mm +/- 0.5 mm could have a measured dimension lying between 100.5 mm and 99.5 mm and the component would still be acceptable.

Attributes are quality indicators that do not allow for variation. They are either acceptable or not, such as a product functioning to specification (or not). For example, a steel oil drum either leaks or it doesn't. If it leaks it is useless and will be rejected. If it does not leak it will do the job it is intended for and will be accepted. Quality indicators can also include attributes that depend on opinion, such as feel, look and appeal. For example, if the feel of a product is satisfactory in the opinion of the quality inspector the product will meet the specification criteria agreed with the customer.

In your coursework project you will need to:

- use **quality control** techniques to manufacture your product
- use quality indicators to check the quality of your product at critical points in its manufacture
- evaluate your product to make sure that it is suitable for your end-user(s).

quality of design and manufacture means that products are well designed and well made at the right price, in order for them to sell successfully at a profit. Quality resistant materials products fulfil their **design and manufacturing specifications** using materials that suit the **end-use**. They fulfil the needs and requirements of **safety legislation**, the **consumer** and the environment.

quality of design means that products are well designed so they sell successfully. For example, a poorly designed child's toy painted in a pastel colour may not sell, even if it is well made. Quality of design refers to a product that:

● is well designed and attractive to the **target market group**

● matches the **design specification**

● uses **materials** that are non-toxic, safe for the environment and that suit the **end-use**

● is easy to manufacture and maintain

● is designed to be suitable and safe for the user and the environment.

quality of manufacture is about making products that are well made so they sell successfully. For example, a badly made child's toy that falls apart after two weeks may have looked attractive when new, but future sales could be limited by its poor reputation. Quality of manufacture refers to a product that:

● is well made, using **materials** suitable for the **end-use**

● matches the **manufacturing specification** and **performance requirements**

● is manufactured by a suitable, safe production method

● is made from durable materials and is easily maintained

● is made within budget limits to sell at an attractive price to the **target market group**

● is manufactured for safe use and disposal, without harm to the environment.

a **quality system** is used by **manufacturers** to control a product's quality. Quality systems incorporate **quality assurance (QA)**, **quality control (QC)**, and **total quality management (TQM)**.

quick response manufacturing (QRM) is a method of production that uses self-organised, multi-skilled teams of employees who are responsible for the **quality** and quantity of product output. QRM is used by **manufacturers** to **batch produce** products quickly in response to **consumer** demand.

QRM has developed through the use of **Information and Communication Technology (ICT)**. This enables fast electronic links, between **retailers** and manufacturers using **Electronic Point of Sale (EPOS)** terminals and ISDN lines.

Product sales information is collected by EPOS terminals. Stock levels are assessed and orders sent electronically to the manufacturer for delivery of fast selling items. Manufacturers make use of **concurrent design and manufacturing (CDM)** and **computer numerical control (CNC)** manufacturing systems to enable fast product manufacture in response to consumer demand. The use of ICT in QRM systems also enables **just in time (JIT)** ordering of raw **materials**, so they arrive just in time for production.

QRM is a cost-effective method of production because it:

- uses an efficient, multi-skilled workforce to increase productivity
- reduces the levels of finished goods put into stock waiting to be sold
- reduces the space needed for keeping goods in stock
- reduces the amount of money tied up in stock.

recycling is becoming a key concern for product **designers** and **manufacturers** because European **legislation** requires environmental protection. Designers need to design with the environment in mind. Manufacturers need to use environmentally friendly processes, waste management techniques and recycle where possible.

Product waste can end up in landfill sites or be incinerated, both of which cause pollution.

- Cellulose (wood fibre) and most metals are degradable so they will, in time, break down in a landfill.
- Synthetic **materials** such as plastics can take more than 100 years to biodegrade and can cause long term land pollution.

The three key approaches to product waste management are 'Reduce, Reuse and Recycle'. Recycling is important because many raw materials, such as plastics, come from non-renewable sources. Most pure metals can be re-melted as cheaply or cheaper than it costs to mine the ores and convert them. The suitability of resistant materials such as wood, metal and plastics for recycling depends on the material type.

In your coursework project you will need to:

- consider the needs of the environment when developing your product **design specification**
- use an efficient layout to reduce waste and cut costs
- use materials and manufacturing processes that reduce risks to the environment
- take account of waste materials when designing and manufacturing.

Source of material	Material type	Recycled material	New product
Window frames, chairs	PVC	Shredded PVC	Garden chairs/tables
Plastic cartons, crates, boxes, bags	Polythene	Shredded polythene	Low cost packaging board
Natural wood products	Cellulose fibres	Paper	Paper products
Ferrous metal products	Iron, steel	New iron or steel	New steel products
Drinks cans	Aluminium	New aluminium	New aluminium products
Non-ferrous metals products	Lead, copper, zinc	New lead, copper or zinc	High cost specialist products such as wiring, circuit boards, flashing for roofing, etc.

Some plastics as well as natural materials, such as wood and most metals, can be recycled

a **retailer** is a person or shop outlet that sells manufactured products. Most retail outlets increase the price of the product by up to 100%. This must be considered when the **manufacturer** costs the product. Retailers like to have products that sell quickly because a fast turnaround increases their profits. They also like products that can be re-ordered at a moment's notice, stack well and take up minimal space. This is another consideration for the **designer** and manufacturer.

risk assessment means identifying the risks to the health and safety (H&S) of people and to the environment. In practice, this means using safe designing and manufacturing processes and making products that are safe to use and safe to dispose of.

Manufacturers use risk assessment techniques to look for possible hazards in their products. They use British Standards to test and monitor production. All possible health hazards to employees have to be eliminated and safety procedures have to be followed to ensure the safety of people at work. Some manufacturers use life cycle assessment (LCA) to assess a product's impact from cradle (raw materials) to grave (disposal).

In your coursework project you will need to:

- include safety criteria in your **specifications**

- take responsibility for recognising hazards in manufacturing processes and work areas

- use safety information to help you assess risks involved in designing, making and using manufactured products

- follow safety rules to reduce risks when designing and making products

- check the safety of your product for your intended user and the environment

- explain how to manage your working environment so you keep risks under control.

Understanding Industrial Practices: Resistant Materials Technology © Nelson Thornes 2004

Design brief	Initial risk assessment: check safety standards, regulations and legislation.
Design specification	Include safety criteria and quality control procedures.
Design ideas	Risk assessment for each idea: check against the design specification.
Manufacturing specification	Include safety criteria, quality control procedures and safety tests.
Prototype	Risk assessment: test prototype against safety criteria in the manufacturing specification and against safety standards. Test in use in extreme conditions.
Manufacturing process	Risk assessment of materials, processes and equipment: use of quality control procedures.
Final product	Risk assessment: test final product against safety criteria in the manufacturing specification and against safety standards. Test in use in extreme conditions.

Risk assessment is carried out at the key stages in the design and manufacture of a product

Coursework checkpoint: *risk assessment*

You need to think about risk assessment when designing and making your product.

1 Draw up a chart to show the key stages in the design and manufacture of your product.

2 For each stage, list the risk assessment procedures required to make your product as safe as possible.

3 Research any appropriate British Standard tests, regulations and **legislation** relating to the design, manufacture, use and disposal of your intended product.

robotics involves the use of computer controlled **machines**. Robots can be programmed and re-programmed to perform continuous and complex manufacturing processes such as:

● pick, place and machine

● manipulative processes such as paint spraying.

Robots can learn their operations from very skilled manual workers who train the robot by just doing their job as normal. For example, an extremely skilled paint sprayer can use a spray gun that has sensors attached to programme the software. Afterwards, a robot arm that holds the spray gun can repeat the exact actions of the paint sprayer.

Robotics enable fast, cost-effective manufacture and can perform repetitive, uncomfortable, dangerous or monotonous processes with greater accuracy for much longer intervals than humans. Robotic machines help keep working environments safe. For example, modern robotic systems automatically shut down if a person is in the way of the operation.

safety in design and manufacturing means the safe design, manufacture, use and disposal of manufactured products. Manufacturers follow **risk assessment** and **safety procedures** and check standards, regulations and **legislation**. This ensures that the products they make are safe for the **consumer** and the environment. Standards and **testing** procedures are set by the **British Standards Institution (BSI)**.

safety procedures are followed to ensure the **safety** of people at work and to prevent industrial accidents. Every risk or source of danger in the workplace has to be identified and safety procedures written down. The table below identifies some of the potential **hazards** in product manufacture. The related safety procedures are listed in the right hand column.

46
47

In your coursework project you will need to:

■ take responsibility for recognising hazards in manufacturing processes and work areas

■ follow safety rules to reduce risk when designing and making your product

■ carry out safety checks on tools, equipment and **machines**

■ work safely with **materials** and equipment when manufacturing your product.

Coursework checkpoint: *safety procedures*

Record in your coursework folder the potential hazards in manufacturing your product. List the safety procedures you need to follow to avoid accidents when using materials and equipment.

Potential hazards	Safety procedures
1 Machining on a lathe	
Finger and body injury from the tool or loose part	Check the guard before work starts. Remove tightening keys or loose parts. Check that all stop switches work properly. Check for loose clothing.
Eye injury from swarf	Check eye guard before work starts. Wear safety goggles.
2 Buffing/grinding	
Hand injury from hot material	Wear safety gloves or wait for the material to cool
Eye injury from flying material	Always use eye guards and wear safety goggles on all spinning machines. Check for loose clothing.
3 Using solvent adhesives	
Hazard from solvent vapours	Make sure that there is adequate ventilation. Clear up spills.
4 General working area	
Tripping and falling	Keep the work area clean, tidy and dry. Beware of trailing cables.
Electrical injury	Never use a machine with damaged cables or switches. Report damage at once. Never try to repair damage yourself.

Safety procedures used in product manufacture

safety wear is designed and made to provide protection in the workplace from flying debris (such as metal swarf or sawdust), heat, cold or chemicals. Workshop safety wear can include overalls, heat-resistant clothes and gloves, boots, helmets, facemasks, goggles and respiratory masks (for breathing). All items of safety wear marketed in Europe have to meet specific performance standards. These are set by the **European Standards Organisation**.

scale of production (method of production, production system, manufacturing system) is decided by the type and number of products to be made. The chosen scale of production enables manufacturers to produce products quickly and efficiently. It can include:

- one-off (low volume or jobbing production)
- batch production
- high volume (mass production, flow-line production)
- continuous production.

sensory tests make use of the human senses of sight, sound, touch, taste or smell to evaluate the aesthetic or ergonomic quality of a product. These tests can be used as quality indicators at critical control points in a product's manufacture.

The sensory qualities of a product such as shape, texture and colour, may affect its performance just as much as accuracy might. For example:

- an effective hammer would have a non-slip grip that easily fits the hand (touch)
- a road safety sign is brightly coloured to attract attention and be read at a distance (sight)
- musical instruments must be capable of being tuned to the correct pitch (sound)
- toothbrushes should have no taste and be non-toxic (taste)
- food processor motors should not give off unpleasant smells when used (smell).

Sensory values can be tested by using a judging panel or by a consumer survey.

size matters to all manufacturers. If a product can be made smaller but perform just as effectively it usually saves on material costs. For example, the cases of many injection moulded plastics products are now designed with all the fixing lugs and partitions as part of the moulding. All the fixing lugs and any partitions are part of the moulding. This reduces the need for separate frames or components and can reduce the size of the product. Smaller products mean that smaller capacity machines can be used for manufacture. Packaging and distribution is cheaper as more products can be delivered at any one time. Smaller products take up less storage space, which is good for both manufacturers and retail outlets.

In your coursework project you will need to:

- take account of aesthetic, **anthropometric** and ergonomic considerations when designing and testing your product

- use a judging panel or consumer survey to test your finished product.

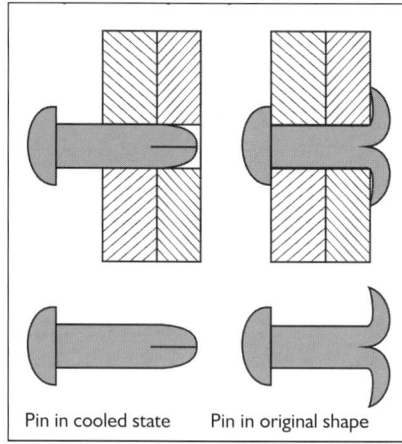

Pin in cooled state   Pin in original shape

SMPs are used to make plastic rivets

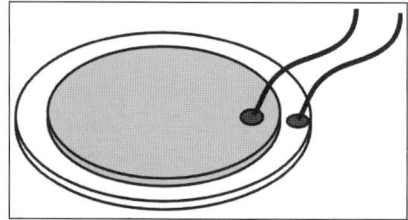

Piezo-electric sensor or actuator
(i.e. tap sensor or sounder)

smart materials respond to differences in temperature, light, moisture, force or electrical current and change in some way. They are called smart because they sense and respond to conditions in their environment. Smart materials appear to 'think' and some have a 'memory' because they can revert back to their original state. For example:

- Thermo-chromatic materials change colour in response to heat, light or moisture. They can be used for sunglasses that get darker in bright sunlight.

- Piezo-electric materials can act as sensors or actuators. If tapped or bent they act as sensors and produce a voltage. For example, they can be used for a spark gas lighter or tap sensor. If a voltage is applied they work as an actuator and bend. For example, as used in a piezo-alarm sounder.

- Flexible materials such as foam can be combined with conductive technology to produce materials that can sense environmental changes such as humidity or moisture.

- Shape memory alloys (SMAs) can be plastically deformed by changing their temperature. When the temperature returns to normal the component reverts to its original shape. For example, 'nitinol' gives a mechanical movement in response to temperature. Other SMAs in the form of coil springs can open and close a hinged greenhouse window in response to changes in temperature.

- Shape memory plastics (SMPs) are used to make components such as special rivets that become straight when cooled, ready for inserting in a hole. When they return to normal temperature the ends bend outwards to work as a fastener rather like a nut on a bolt.

- The simplest example of a plastic with a heat memory is acrylic. When heated and bent, then cooled, it retains the required bent shape. When reheated it returns to its original shape. However once this has happened it doesn't change its shape again when reheated or cooled so acrylic is not a true SMP.

specifications are the design and manufacturing criteria that are used to determine and monitor the **quality** of manufactured products.

A product **design specification** guides a **designer's** thinking about what is to be designed. It is used to guide research, to test and evaluate design ideas and to develop the **manufacturing specification**.

A manufacturing specification ensures that a product is manufactured as the designer intends. It forms part of the production plan for the manufacture of the product. The manufacturing specification is used as a standard for checking the quality of each product.

In your coursework project you will need to:

- develop and use detailed design specification criteria

- use your design specification to generate, test and evaluate design ideas

- use your manufacturing specification to evaluate the quality of your product.

Understanding Industrial Practices: Resistant Materials Technology © Nelson Thornes 2004

Design specification	Manufacturing specification
Developed from the design brief, research and analysis.	Developed from the design specification and the final solution
Specifies the design limits	Specifies the product manufacturing limits
Used to generate, test and evaluate design ideas	Used to test and evaluate the product after manufacture
Used to monitor the quality of design	Used to monitor the quality of manufacture
Used to develop a manufacturing specification	Used to develop a high quality finished product
Specifies the proposed characteristics of the product	Specifies the working characteristics of the finished product

The differences between a design specification and a manufacturing specification

the **stages of production** for most manufactured products include preparation, processing, **assembly**, **finishing** and packaging.

Preparation can include:

- buying-in of **materials** and components
- preparation of materials, components, tools, equipment and machinery
- the production of **working drawings** and **layout plans**, **jigs** and **templates**.

Processing can include:

- cutting of component parts
- drilling, turning, cutting or abrading, moulding, forming, casting
- finishing **sub-assemblies**, e.g. painting sections of a wooden toy before final assembly.

Assembly can include:

- sub-assembly, e.g. assembling a working mechanism
- **joining** components or sub-assemblies, using welding or soldering, mechanical fastenings or adhesives.

Finishing can include:

- decorative finishing, such as painting to improve the appearance of the product
- **functional** finishing, such as plating, coating or sealing to improve the product's **quality** and resistance to the environment.

Packaging can include:

- **labelling** with labels, tags and **bar codes** to identify product type, price information and stock details
- packing in an outer layer to protect the product
- packing in boxes for easy transport
- features that help to sell the product such as a photo, the manufacturer's logo, **brand** name and maintenance instructions.

> **In your coursework project you will need to:**
>
> - produce and use a **work schedule** that shows the stages of production for your product.

In your coursework project you will need to:

■ consider the use of standard components to make your manufacture more cost-effective

■ choose standard components that are quick and easy to use.

standard components include screws, hinges, locks, electrical fittings, motors, lamps, clock mechanisms, knock-down fittings and other components that can be bought off the shelf. Standard components are widely used to reduce the cost of manufacturing products because they:

- are readily available from a range of suppliers
- are available in known standard **sizes** that enable cost-efficient manufacture
- provide a range of high **quality** component parts at low unit cost
- can be bought in bulk at a reduced unit cost
- can be used in product costing, because their costs are known
- can be ordered **just in time** (JIT) for delivery when they are needed
- reduce the need for component sub-assembly at the **manufacturers**.

Some standard components, such as knock-down fittings, can be used to further reduce manufacturing costs because:

- they make it possible for **consumers** to assemble the final product
- they reduce storage and transportation costs because products take up less space and can be delivered as flat-packs.

a **structural frame** is a framework assembled from an arrangement of struts and supports which are either in compression or tension. They are designed to resist known forces at the lowest cost. 3D structural frames are used for **architectural** constructions, bridges and for machines such as cranes. They are also used for smaller products that need to be strong but lightweight such as bicycles, climbing frames, tent supports, chairs, tables and sit-on toys. **Designers** develop structural frames to reduce the weight of **constructions** and keep costs down by using the least amount of material for the greatest strength.

Effective structural frames should:

- be stable and lightweight
- be strong enough to resist known forces without breaking
- be economical in their use of **materials**.

In your coursework project you will need to:

■ recognise that cultural, social and environmental considerations can influence the development of product style

■ use the influence of past styles and the work of other designers when developing product ideas.

style refers to the look of a product. It depends on product purpose, the **colour**, **material** and trends at a particular time. The **style** that is used by **designers** depends on the **target market group** and their expectations. The design and style of many products is influenced by fashion trends. Many different words are used to describe style, such as:

- classic, e.g. simple but elegant
- feminine, e.g. softer shapes and colours
- geometric, e.g. rectangular, circular, triangular
- masculine, e.g. angular, stronger colours
- modern, e.g. using new technologies or materials, current trends
- rustic, e.g. country styles, unrefined, simple
- sporty, e.g. streamlined, bright colours
- traditional, e.g. past styles.

a **sub-assembly** is a section of a product that has to be put together in a pre-determined order so that the product can be completed. For example, a product that has a mechanism usually needs this assembled first or in order, otherwise it may not fit or work properly.

A sub-assembly is also a section or complete part of a larger product. For example, the braking system for a bicycle could be called a sub-assembly as it is a complete working section that can be made and tested on its own. However, it is just one part of the whole bicycle. Many products use ready made sub-assemblies from other manufacturers. For example, it would be uneconomic for the manufacure of batch produced wooden clocks to make clock mechanisms. Specialist manufacturers can make the mechanisms in great quantities at very low cost, which enables them to be bought in on a just in time (JIT) basis.

a **swipe** is a picture cut from a magazine. Swipes are used on mood boards to inspire and develop ideas about colour and style.

a **system** is a co-ordinated arrangement of activities working together in which inputs are processed to achieve outputs.

In a simple manufacturing system the INPUT is the raw material, the PROCESS is manufacturing the product and the OUTPUT is a finished product delivered to the retailer. This way of thinking about activities is called a 'systems approach' as the system can be broken down into manageable parts that can be easily organised. A simple manufacturing system can be represented by a systems diagram (block diagram, flow diagram), which shows how the system works. In a simple kind of system there may be no feedback of information so it is called an 'open loop' system.

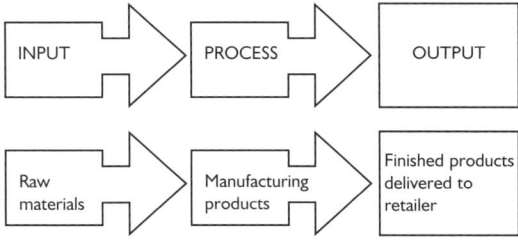

Block diagram to show a simple 'open loop' manufacturing system

Manufacturing systems that work well and result in high product sales, incorporate feedback of information. This is called a 'closed loop' system. In quick response manufacturing (QRM), for example, Information and Communication Technology (ICT) systems allow the feedback of product sales information from Electronic Point of Sale (EPOS) tills to the product manufacturer. This feedback of information enables the manufacturer to plan production and manufacture goods to order.

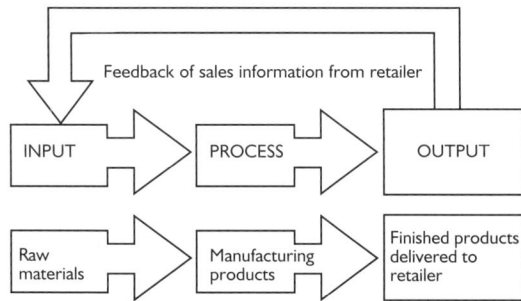

Block diagram to show a 'closed loop' manufacturing system

Understanding Industrial Practices: Resistant Materials Technology © Nelson Thornes 2004

In your coursework project you will need to:

■ understand how systems are used in industry to control design and manufacture

■ understand that systems are made up of inputs, processes and outputs and that feedback is used to make a system work well

■ use a **quality control** system during manufacture that incorporates feedback so your product is fault-free

■ use feedback from users to help you improve your product.

Feedback is also used in manufacturing systems to improve reliability and **safety** during production. For example, feedback in a process checking routine can automatically stop a machine from working if dangerous limits are exceeded, such as too high temperatures in a plastics moulding process.

Large manufacturing systems are often made up of smaller sub-systems or sections, such as product design, **assembly**, finishing, packaging, stock control, sales and **marketing**. These can be linked together to make the whole system work in a cost-effective way. This linking is possible through the use of **product data management (PDM)** software, to enable **computer integrated manufacture (CIM)** and **concurrent design and manufacture (CDM)**.

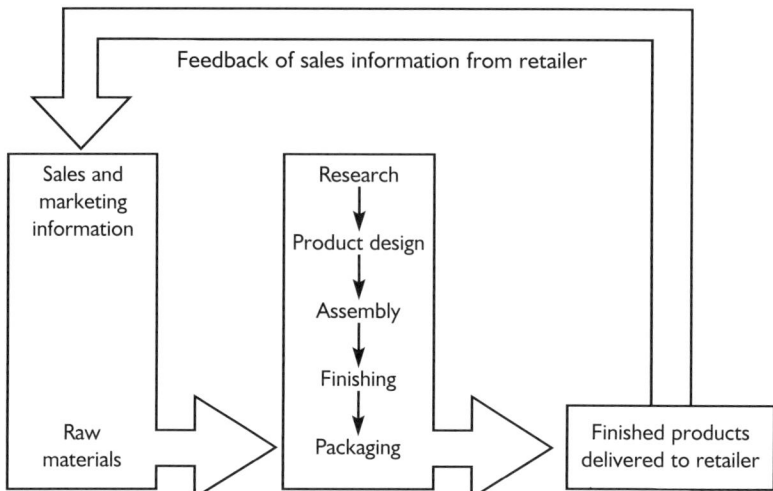

Block diagram to show a large manufacturing system with sub-systems

Types of systems used in product manufacturing include:

- **control systems**
- **costing** systems
- ICT systems
- manufacturing systems
- marketing systems
- PDM systems
- **production systems**
- **quality** systems or **quality assurance** systems
- sales and distribution systems
- safety systems.

systems and control is a term used to explain how **systems**, control systems, computer systems, quality systems and production systems are used in manufacturing. Different types of systems and control are used to monitor the whole manufacturing process so that products are manufactured efficiently at a profit.

the **target market group** is the consumer group that a manufacturer aims to sell to. Manufacturers use market research to identify the requirements and size of the target market group so that their products will reflect the needs of the consumer. Manufacturers identify their target market groups according to age, quality needs, leisure activities, lifestyle or style awareness.

teamwork in manufacturing is a way of achieving greater flexibility of production, a faster time to market and improved working conditions for employees. It has been defined by the Centre for Work and Technology as: 'a flexible, quick response system, consisting of self-organised, self-motivated, multi-skilled, versatile personnel who work collectively in teams, making joint decisions and sharing responsibility for output in terms of both quality and quantity'.

In manufacturing industry teamwork can take different forms:

- In a production line arrangement each team member performs one or more assembly processes repeatedly.
- In quick response manufacture (QRM) each team member works in a production cell to produce complete products. Each production cell is responsible for its own quality control as the product is made. This speeds up production and makes each worker feel more valued for their overall performance.

Many product manufacturers use cell manufacture for batch and high volume production because teamwork provides benefits to the manufacturer, the workforce and the consumer.

1 Benefits of teamwork to the manufacturer are:
- a flexible, multi-skilled workforce
- increased productivity and efficiency
- retained orders from retailers, through reliability and QRM
- improved quality, through team responsibility
- reduced absenteeism, through greater commitment to work
- reduced labour turnover, through improved working conditions.

2 Benefits of teamwork to the workforce are:
- improved working conditions and increased motivation
- opportunities for decision-making, to solve problems
- greater variety so less boredom
- increased average earnings
- improved relations with management.

3 Benefits of teamwork to the consumer are:
- improved product quality, through team responsibility
- improved delivery performance, due to speed of manufacture
- fast response to consumer demand, through flexibility
- fast changeover for new styles
- wider variety of goods available.

In your coursework project you will need to:

- take account of a range of users in your target market group

---

In your coursework project you will need to:

- choose the most suitable **method of production** for your product and give reasons for choosing it
- understand why QRM and teamwork improve product quality and output
- simulate production and assembly lines.

In your coursework project you will need to:

- design and use quality checks to test the fitness for purpose of your product

- check the quality of your product at critical points in its manufacture

- test your product to make sure that it is suitable for its **end-use**

- suggest ways of improving the quality of your product.

testing of materials and components before manufacture is part of a quality assurance (QA) system that requires the use of standard tests under controlled conditions. Standard performance tests can be set by the British Standards Institution (BSI) and by individual retailers.

Checks on prototypes are made to test for performance, ease of manufacture, maintenance and fitness for purpose. These tests ensure the production of quality products, avoid costly mistakes and protect the consumer against faulty or unsafe goods. In industry tests make use of:

- manufacturing specifications to test products against tolerance limits and quality standards

- manufacturing prototypes to test for ease of manufacture and maintenance

- consumer testing to check buying trends and reactions to new products. Sometimes consumers use products in a way they weren't intended. This may raise safety issues which need feeding back into the design development.

Coursework checkpoint 1: *testing*

1 In your coursework folder only include testing that is directly related to the development of your product. For example, if your product is a wooden garden seat you need to test for seating height, stability and for **finishes** that are resistance to weathering. If it is a jewellery container tests for **size** and **ergonomic** efficiency are less important but how it looks may be critical. Look up the expected **properties** of your **materials**, **joints** and finishes to explain why they are suitable for your product.

2 Only experiment with techniques you have decided to use (such as mortise or dowel joints to test for comparative strength). This will enable you to develop your skills properly using the appropriate techniques.

3 In your coursework folder only include descriptions or explanations of the actual techniques and processes you use to develop your product. Your knowledge and understanding of other processes will be tested in your written examination.

Coursework checkpoint 2: *testing*

Products and processes need to be tested before and during manufacture to check their fitness for purpose and to save time and costly mistakes later on.

Design a series of tests that are linked to your product specifications. This will help you produce a high quality, valued product. You could design tests or quality control checks for:

- the suitability of the process (e.g. strength of joints and ease of manufacture)
- the suitability and cost of fastenings
- the size of the product and its component parts
- achieving tolerance limits and quality standards during manufacture
- an efficient and safe method of **assembly**
- product performance
- **product maintenance**
- the safe use of the product
- consumer reaction (to the product **style**, **colour**, shape, cost).

tolerances are allowable variations in the dimensions of a manufactured product to ensure it fits together accurately and meets all the manufacturing specification criteria. These allowable variations in dimensions can be included in the manufacturing specifications either on a working drawing or in a work schedule.

Tolerances enable the component parts of the product to be measured for accuracy against defined limits. They are normally only given:

- for component parts that have to fit together accurately
- for holes, slots and locations for components
- where dimensions are critical to the function of the product.

For example, a joint dimension with a tolerance of 25 mm 0/–0.5 mm would mean that joints could be cut exactly to size or with a variation of up to –0.5 mm (i.e. as small as 24.5 mm) and the product would still fit together well. However, if the joints had a tolerance of 25 mm +/–0.5 mm it could mean that some joints would be too tight as they may be +0.5 mm oversize (i.e. as large as 25.5 mm).

Some tolerance limits can be much wider. For example, the tolerance limit for the height of a garden seat could be as much as +/–25 mm and the seat could still function well. Some tolerance limits can be much smaller. For example, the tolerance limit for a metal bearing used for an axle may need to be less than 0.1 mm for the axle to work well without wobbling. Critical tolerances are always given where fits or dimensions make a difference to product operation or quality.

The table below shows how tolerance can be defined for a component part. The tolerance limit for a plain bearing is defined for the outside and internal diameters, which are critical dimensions, but the length of the plain bearing has no tolerance specified because it is not a critical dimension.

Part	No.	Length mm	Outside diameter (OD) mm	Internal diameter (ID) mm	Material
Bearing	4	15	13 +/-0.1	10 +0.1	Bronze

Tolerance levels for a plain bearing

Too many variations from defined tolerances will produce faults, resulting in a product or component not fitting together accurately or working within its specifications. Quality control procedures would cause the rejection of the product or component.

total quality management (TQM) is a planning process used in a quality system to make sure that manufactured products fulfil agreed quality standards. The basic idea behind TQM is to produce products that are right first time.

Most industries that wish to achieve TQM also try to introduce automated inspection processes to improve quality. Additionally, there is a move towards computer integrated manufacturing (CIM) systems of manufacture that aim to reduce human input and guarantee reliability and quality control. To achieve this, the workforce needs to be more skilled at setting up and maintaining the necessary computer operated equipment, which in some circumstances can run continuously as long as a maintenance and servicing schedule is included.

In your coursework project you will need to:

- define tolerance limits in manufacturing specifications and working drawings

- produce a quality control system to monitor accuracy and tolerance so your product meets manufacturing specifications

- work accurately when making your products.

TQM aims at continuous improvement, not only in meeting product specifications, but also in the training and well-being of its workforce. People who feel valued generally do a better job.

a **trade mark** (brand name) is used by manufacturers to protect and promote a manufactured product or process. This ensures that it cannot be legally copied by competitors.

user trials are tests in which products are tested for quality, performance in use, safety and sales appeal, usually by consumer panels. The results of user trials are used to modify, where necessary, the design or manufacture of the product.

Many manufacturers use British Standards tests and quality assurance (QA) systems to make sure that their products fulfil the safety and quality needs of consumers.

> **Coursework checkpoint:** *user trials*
>
> 1 You can trial your product during manufacture by regularly testing it against the **manufacturing specification** to check each part works properly.
>
> 2 After manufacture, a user trial of the product can provide valuable information about the product's performance in use. It can form part of your product evaluation.

a **user trip** is a way of letting a designer feel like the consumer the product is designed for. For example, if a wheelchair is being designed it is helpful to experience the sort of problems real users face by being in one for a day. If a child's pushchair is being designed it is helpful for the designer to push one around a town, preferably with a child in it, do some shopping and get on and off public transport to experience the real problems parents face.

values issues include cultural, moral, social, economic and environmental considerations, which have a very important influence on the design and manufacture of manufactured products. Manufacturers must take into account the needs and wants of the market they are selling to. Products that are perceived by the target market as 'must haves' sell well and make a profit for the manufacturer.

Cultural and social values relate to fashion and lifestyle. Trends, fashion and style influence the development of many different types of product, such as furniture, sports equipment, accessories, e.g. sunglasses, domestic products and cars. Social influences from the media, films, music, TV and exhibitions also have an impact on product design.

Values issues are important because products carry powerful messages about the people who use them. For example, products such

55

In your coursework project you will need to:

■ use tests, modifications and evaluation to make sure that your product is suitable for its **end-use.**

In your coursework project you will need to:

■ make good use of social, cultural and environmental influences when developing product ideas

■ use the influence of past times and the work of other designers when developing ideas.

as 'fashion' watches or sunglasses are often chosen for a required 'image'. These images are often related to 'brands' that are perceived as 'special' in some way. The status and image of a strong **brand** and brand loyalty can therefore have a powerful influence over which products people buy. **Designers** have to consider very carefully the values, **needs and wants** of the market at which the product is aimed.

Values issues can also be the 'driving force' behind product development. For example, environmental **legislation** has led to the development of new **materials** and products, such as environmentally friendly biodegradable polythene, which is used for bags and other food containers. Another consideration in product development is the moral dilemma of **global manufacturing** – designing in Europe and manufacturing overseas. Although global manufacturing has resulted in the availability of inexpensive products, there may be a moral 'price' to pay in terms of where and how these products are manufactured.

Coursework checkpoint: *values issues*

Think about some of the following issues when you are developing your product:

- Why some products are more popular than others.
- Why there are 'crazes' for some products.
- The role of brands in selling products.
- The impact of peer group pressure on the products people buy.
- The influence of 'celebrities' or TV on fashion trends.
- Buying cheaper illegal copy products in the market.
- Buying inexpensive fashion items, such as sunglasses, from supermarket chains. Where are these products made?

---

virtual products can be developed on screen using computer software. This type of product development uses **computer-aided design (CAD)** software to design and show products on screen, usually from different angles and fully rendered by being shaded or coloured.

Designing virtual products can save a **manufacturer** time and money because it reduces the need to produce real product samples. A simulated product can be generated on a computer screen where it is relatively easy to change the **colours**, proportions and dimensions once the initial **modelling** is complete. It is also quite easy to show the product in a simulated real environment by using a backdrop. For example, skis and other winter-based products can be shown in snow-covered mountains without actually having to take them there.

the **visual appearance** of a manufactured product can be used as a sensory test to evaluate the **quality** of the product. Visual appearance is a **quality indicator** used as a **quality control** check at **critical control points** in the manufacturing process. In the final quality check, the visual appearance or 'look' of the product is compared with a perfect sample.

a **work schedule** (production schedule, manufacturing schedule) is the written breakdown of the production processes needed to manufacture a set quantity of components or products. It gives details about the order of production, the equipment required, the process time and in some cases the **tolerances**. The aim of the work schedule is to make sure that each product is made to the same **quality** so it is manufactured right-first-time-on-time.

The work schedule is a key part of the production plan and **quality assurance (QA) system**. It identifies:

- the processes needed to manufacture a set quantity of products
- the order of production
- the time each process will take in seconds or minutes
- the machinery or equipment required
- dimensions or tolerance limits or where they can be found (usually on **working drawings**).

The work schedule or working drawing records tolerances that are used to check product quality during manufacture. Any faults in the making are identified and action is taken to correct the fault. For example, if a moulding machine is running too hot or too cold it will seriously affect the quality of the product. Too hot may result in the plastic parts either taking too long to cool which wastes production time, or worse the plastic may distort and the part become useless. Too cold may mean the plastic doesn't fill the mould in the first place and so the part has to be rejected. A greater problem still is that the machine could be damaged and there is also risk of fire. Any changes made are recorded in the production plan to prevent the fault happening again and to enable the manufacture of successful identical products.

56

Work schedule: P125D for 5 clock faces
See CAD working drawing on worksheet 4 for dimensions and critical tolerances

Date 08 Nov	No.	Size 125mm	Style P125D		
Order of process		**Material**	**Machine**	**Process**	**Process time (seconds)**
1	1	Walnut	Circular saw	Cutting 700 × 130 × 10 strip	120
2			CNC router	Secure strip in holding jig	60
3	5	Walnut	CNC router	Cutting centre holes	20
4	20	Walnut	CNC router	Cutting hour indicator holes	100
5	40	Walnut	CNC router	Cutting hour indicator holes	200
6	5	Walnut	CNC router	Cutting perimeter shape	300
7	5	Walnut	Belt sander	Finish sand (edge and face)	300
	5	Walnut			
				Total process time	1200 (20 mins)

*Part of a simple work schedule used for the batch production of **5** wooden clock faces. See working drawing of clock 1 on worksheet 4 for dimensions and critical tolerances.*

In industry the work schedule can record quite detailed information about the time needed to make a set quantity of products. For example, the table below shows the time needed to manufacture 200 wooden boxes. Once the total production time is worked out, a **Gantt chart** is drawn up to schedule the manufacture of the boxes on a daily basis. A **production planning** form is also drawn up to record details of resource requirements, such as human, tooling and materials.

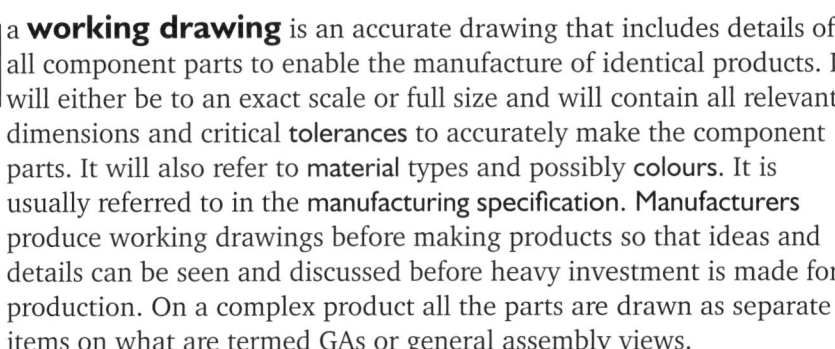

Order of production	Machine set-up time (mins)	Number of operators	Number of parts	Time per operation (mins)	Production time for 200 boxes (mins)	Machine set-up & production time (mins)
Cut base	2.00	1	200	0.25	200 x 0.25 = 50	52
Cut ends	2.00	1	400	0.25	100	102
Cut sides	2.00	1	400	0.25	100	102
Cut base grooves	5.00	1	400	0.20	80	85
Glue and tack		2	200 sets	3.00	600	600
Sand smooth		2	200 assemblies	2.50	500	500
Paint		2	200 assemblies	3.00	600	600
Total production time to manufacture 200 wooden boxes						2041 mins (34 hours 1 min)

Coursework checkpoint: *work schedule*

Remember, your work schedule is *part* of your production plan that also includes your **manufacturing specification**, working drawings and a **flow diagram** to show where and how to check for product quality.

Remember to record any changes you make when assembling your product. This will enable you to explain how you could manufacture identical products in **high volume**.

a **working drawing** is an accurate drawing that includes details of all component parts to enable the manufacture of identical products. It will either be to an exact scale or full size and will contain all relevant dimensions and critical **tolerances** to accurately make the component parts. It will also refer to **material** types and possibly **colours**. It is usually referred to in the **manufacturing specification**. Manufacturers produce working drawings before making products so that ideas and details can be seen and discussed before heavy investment is made for production. On a complex product all the parts are drawn as separate items on what are termed GAs or general assembly views.

Many working drawings are now produced using computers and **computer-aided design (CAD)** software. This enables rapid changes and updates to be made. The data produced can be used to drive **computer numerically controlled (CNC)** machines to make the product or its component parts.

The working drawing shows:

- detailed dimensioned views of the product and its component parts (usually front, side and plan views)
- the design and styling details including **finish**
- details of fastenings, fixtures or special features such as mechanisms that need to move

57

In your coursework project you will need to:

- use hand techniques or computer software to produce accurate working drawings
- show accurate dimensions and critical tolerances
- take account of the degree of accuracy needed when planning your product manufacture
- decide on a range of techniques and processes to make your product.

- material types and possibly colour (plastic products will need colours)
- material sizes in a component **parts list**
- dimensions and critical tolerances.

This information enables a **prototype** to be made to the specified dimensions using the intended materials and similar processes to the ones to be used for the manufactured product. If the prototype works well, production tooling is made and used to produce a sample product run before full-scale production begins. If any modifications are required to the product new working drawings have to be produced.

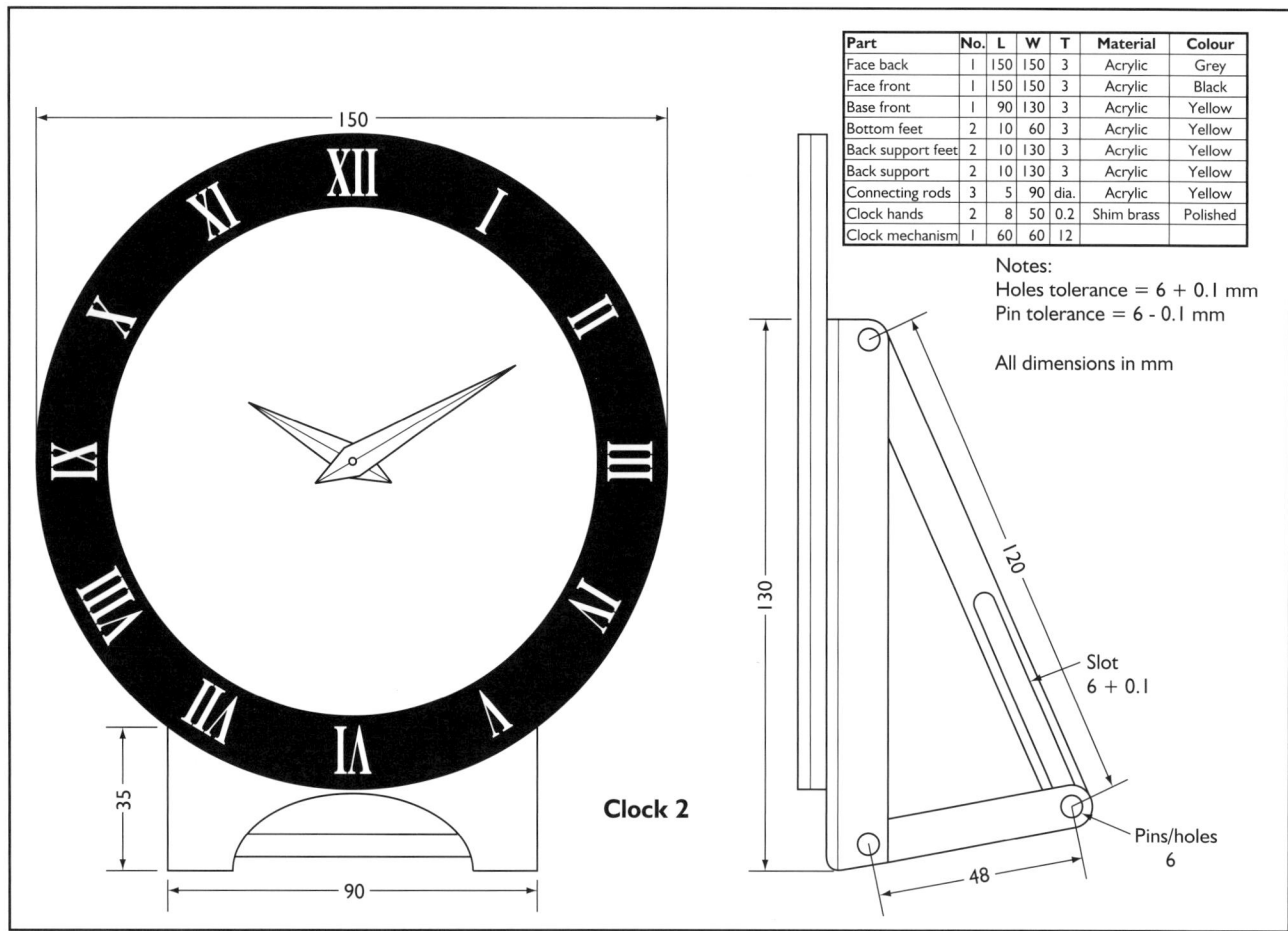

Part	No.	L	W	T	Material	Colour
Face back	1	150	150	3	Acrylic	Grey
Face front	1	150	150	3	Acrylic	Black
Base front	1	90	130	3	Acrylic	Yellow
Bottom feet	2	10	60	3	Acrylic	Yellow
Back support feet	2	10	130	3	Acrylic	Yellow
Back support	2	10	130	3	Acrylic	Yellow
Connecting rods	3	5	90	dia.	Acrylic	Yellow
Clock hands	2	8	50	0.2	Shim brass	Polished
Clock mechanism	1	60	60	12		

Notes:
Holes tolerance = 6 + 0.1 mm
Pin tolerance = 6 − 0.1 mm

All dimensions in mm

Clock 2

Working drawing of a clock

Coursework checklists

1 How to manage your coursework project

- Find out the project deadline so you know how many weeks you have for your coursework.
- Read the assessment criteria so you know how research, design, manufacture and evaluation will be assessed.
- Produce a **Gantt chart** for your project so you plan your time well.
- Find out the number of pages you need to have in your project folder.
- Too much research can be a waste of time. Ask your teacher how much research you should do.
- Tick off your teacher's checklist as you work through your project.
- Try grading your own project against the assessment criteria.

2 What to include in your coursework project folder

When designing you should include:

- a **design brief**, stating the type and purpose of product you will design and the **target market group** it is aimed at.
- an analysis of the design brief resulting in a list or word map to focus what you need to research. You can use a **design theme** to help your research. Check your research list with your teacher.
- a research analysis and summary of commercial products similar to your own with information about **materials**, processes, **performance requirements**, value for money, **product maintenance**, **quality** and **safety**. Also include the needs and values of the target market group and current design trends.
- only research information that is relevant to the development of your own product. Check your research analysis and summary with your teacher.
- a product **design specification** that sets out detailed criteria about the product you will design. Make sure that you use your design specification to guide your design ideas.
- a brainstorm, mind map, word map or annotated **mood board** to explore and explain your first ideas about your product. Use this as a starting point for generating design proposals.
- a number of design proposals that are evaluated against your design specification criteria.
- a design solution that is developed from one or more of your design proposals. Your design solution should meet your product design specification criteria.

When developing a final solution and planning for production and manufacturing you should include:

- clear evidence of **modelling**, **prototyping** and **testing** your design solution. Allow enough time to evaluate your work as it progresses. Modify your design solution if necessary. Record and explain any changes you make.
- a **manufacturing specification**, production plan or **work schedule** that details how to manufacture one product and how it could be manufactured in quantity.
- clear evidence that safety measures were put in place and acted on during the manufacture of a high quality product that meets your design and manufacturing criteria.
- evidence of evaluation against **specifications** throughout the design and manufacturing process. Compare your product with a similar commercial one and record the views of the target market. Suggest how to improve the marketability of your product.

3 How to show evidence of industrial practices in your coursework

- Use the **key words** (see page 79) to help you look up and use technical terms similar to those used in industry.
- Use the A–Z and worksheets to help you use designing and manufacturing activities similar to those used in industry.

4 How to use ICT in your coursework

- At the start of your project, make a list of how you could use Information and Communication Technology (ICT) and computer-aided design (CAD) for research, designing, modelling, communicating and testing.
- Make a list of how you could use ICT, including CAD/CAM for analysing information and production planning to support the manufacture of your project.

5 How to use systems and control in your coursework

- Plan how you will use computer systems to support the designing of your product.
- Plan how you could use a CAD/CAM system to control the manufacture of your product.
- Plan how you will use a quality control system that incorporates feedback to help you manufacture a high quality product.
- Plan how you will apply safety procedures in the manufacture and use of your product.

Key words

Context for design

- Design brief, situation/context for design, problem, needs and wants, product performance, end-use, scale of production, time limits.
- Target market group, clients, designers, manufacturers, retailers, consumers, users.

Research

- Diary, log, notebook, record.
- Product analysis, aesthetic properties, trends, market research, design theme, product cost and quality.
- Target market group, user needs, size and shape, anthropometric data, ergonomics, fashion, lifestyle, maintenance, environmental issues, recycling, values issues.
- Available resources, materials, fastenings, standard components, processes, health & safety (H&S), British Standards.
- Internet, Information and Communication Technology (ICT), questionnaire, shop report, user trip.

Analysis

- Analysis, conclusions and summary of research.
- Values issues.

Product design specification

- Design specification, design criteria.
- Clients, designers, manufacturers, consumers/users.
- Product end-use, function or purpose.
- Product users, target market group.
- Aesthetic properties, performance requirements, materials, standard components, fastenings, processes, scale of production, product life, cost, maintenance.
- Quality requirements, cost limits.
- Safety requirements, labelling, legislation, British Standards.
- Environmental and values issues.

Design ideas

- Brainstorm, mind map, annotated mood board, swipe.
- First ideas, creative, design proposals, sketches, annotation, available resources, colour, evaluation against design specification criteria.

Design development

- Design solutions, annotation, colour, materials, processes, CAD/CAM, evaluation against design specification criteria, modelling, modification of solution.

Materials

- Specific types of wood, metals and plastics.
- Smart materials, composites.

Aesthetic and functional properties of materials

- Appearance, colour, grain, texture.
- Durability, strength, hardness, weight, flexibility, stability, corrosion, fire resistance, weather resistance.

Modelling and prototyping

- computer-aided design (CAD), 2D modelling, 3D prototyping, testing, virtual product, processes, costing, spreadsheet.
- Time to evaluate, modify, fitness for purpose, record and explain changes.

Planning production

- Manufacturing specification, designing for manufacture, computer-aided design (CAD), working drawings, production plan, sequence, work schedule, costing, jigs and templates, sizes, tolerances, fit.
- Gantt chart, flow diagram, input-process-output, just in time (JIT), quality checks, health & safety (H&S), risk assessment, hazards.
- Production system, computer system, control system, quality assurance (QA), safety system, risk assessment, total quality management (TQM), one-off, batch production, high volume production, continuous production, quick response manufacturing (QRM), automation, computer integrated manufacture (CIM).

Manufacturing

- Processes, tolerances, computer-aided design (CAD), working drawings, jigs and templates.
- Computer-aided manufacture (CAM), computer numerical control (CNC), jigs and templates.
- Stages of production, cutting, machining, joining, fabricating, assembly, finishing.

Machines and processes (manual and CNC)

- Lathes, mills, drills, routers, planers, moulders (injection/vacuum/blow/compression), extruders, strip heaters, grinders, sanders, buffers, saws, laser cutters, vinyl cutter/plotter.
- CAD/CAM, computer integrated manufacture (CIM), robotics.
- Decorative techniques, finishes, spraying, painting, coating, plating.

Evaluation and testing

- Quality control, critical control point (CCP), quality indicators, feedback, inspection, specifications, tolerances, product performance, fitness for purpose.
- Target market group, user trial, feedback.
- Modifications, improvements, marketing.
- British Standards Institution (BSI), labelling, legislation.
- Diary, log, notebook, record.

1   anthropometrics

Getting your product to be the right size, **colour** and feel for your **target market group** will help your products to be more successful.

1 You have been asked to design a domestic telephone for use by people with poor grip or co-ordination. They will have difficulty using a standard telephone, as the buttons may be too close together.

- Measure and draw full size the buttons for a standard telephone.

- Make a table or chart of finger sizes. You will need to record index finger length and thickness.

- Work out the smallest button and gap that should be suitable for nine out of ten people.

- Try putting thick gloves on and operating a telephone yourself with the hand you do not write with. Describe what effect this would have on the smallest suitable button and gap. Estimate the change needed to your final button and gap dimensions.

- Print the numbers 0–9, * and # as they are on a telephone keypad. Mount them on to thick card and cut them out to the sizes you have calculated to be large enough to work with thick gloves. Make up a keypad with the new buttons and retest the sizes of the buttons with your gloves on. Record your results and make further modifications if necessary.

2 Most seats and tables that are designed for written work are at the same height. This is because it is more expensive to make the heights adjustable or provide a range of different heights.

- Using boxes, cushions or other suitable packing materials, work out what would be the best sitting height for you to write at a classroom table.

- Using the same method, take the same sitting height measurements for the tallest and smallest person in your class and work out the best height to suit most students. Would you choose to use the highest or the average sitting height for working at a table?

3 You are asked to design a sit-on, push-along toy trike for children aged 3–6 years old.

- Take and record leg measurements from children aged 3–6 years old. You may be able to use ready-made charts for this but it is useful to take actual measurements if you can.

- Take and record the measurements from the seat to the middle of the back so that a back support could be designed for the trike.

- Take and record measurements for reach so that handles could be provided at the correct distance.

- Determine the best ride height for the sit-on, push-along toy trike.

Further work

For one of the tasks listed above, produce a range of product ideas on an A3 sheet.

Understanding Industrial Practices: Resistant Materials Technology © Nelson Thornes 2004

2　assembly

assembly

Name . . . .

Form/group . . . . **Date** . . . .

If you look closely at a resistant materials product like a tape cassette, you will see that it is made from a number of separate component parts.

1　Using a separate sheet, sketch each component part of the tape cassette. How many parts are there? Which parts are the same shape and size?

2　Number the component parts in the order or sequence you think they need to be assembled. Using the **work schedule** below carry out the following:
   - list the component parts in the order of **assembly** (the first one is given)
   - list the assembly process (for example, clipped, screwed, heat-welded, pressed, located)
   - name the type of **machine** that would be used for automated assembly.

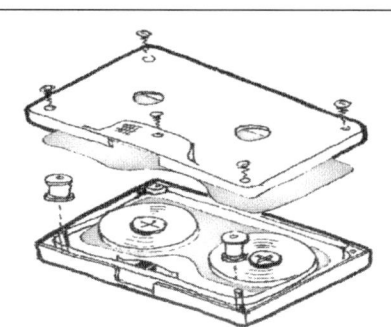

Parts: bottom + top case halves, screws, 2 spindles + tape guides, bottom + top low-friction liners, tape spools, central tape pressure pad

When you write a work schedule for your own coursework project remember to include the time taken for each process in the column headed 'Process time'.

Work schedule: Cassette case (Assembly)				**Date** 18 Nov	
Note: pick/place parts location tolerance for CNC robot: +/– 0.1 mm					
Order	**No.**	**Component part**	**Process/assembly** (place, clip, screw, weld, fit)	**Machine**	**Process time**
1		Bottom case half	Place	CNC robot	
2					
3					
4					
5					
6					
7					
8					

Further work

Produce a simple work schedule for a clothes peg, toothbrush, paper glue stick, stapler or other simple item with few component parts.

3   batch production

Working as a team, design and make a batch of high **quality** mats for standing drinks on. You could design the drinks mats using wood or plastic materials. You will need to consider making **jigs** or **templates** to make sure that each product is made the same **size** and shape. Keep the design simple so the products can be manufactured quickly and efficiently. You could design and make several batches of drinks mats to sell at the school fair.

You will need to:
- set up a design and production team of 3 to 6 students
- allocate and agree each team member's role. The following list of design and manufacturing activities will help you agree on **teamwork** roles.

Design and manufacturing activities
1 The whole team researches **consumer** preferences and decorative techniques regarding **style**, image and **colour**.
2 Each team member designs one solution and produces a **prototype** product (possibly in card).
3 The whole team evaluates each prototype and one is chosen for production.
4 The whole team discusses the production plan and manufacturing process. Each team member agrees to do one manufacturing job such as the production manager, quality controller or technician (more than one person could be a technician).
5 Batches of drinks mats are made using the production plan, bearing in mind **quality control**.

Job descriptions
- The production manager works out the easiest way to make the batch of drinks mats, checks that material and equipment is available and works out the product manufacturing cost. The selling price needs to be high enough to make a profit yet be attractive to consumers.
- The quality controller (QC) identifies where and how to check for quality to ensure each product is made the same. This needs to be done during as well as after production to check the product meets consumer expectations.
- The materials technicians/engineers cut the materials and make and decorate each mat.

Team role	Name of student
Market research	
Design/prototype	
Production manager	
Quality controller	
Technician	

Decide what each person in your team will be responsible for.

Further work

1 Evaluate how successful your team was in designing and **batch producing** your product.
   - How did you assess the quality of your product?
   - How could you improve the design and the production plan?
   - Evaluate how well each team member worked. Did anyone have enough or too little to do? How could you improve on this? Make notes to explain this.
2 Plan how you could market your batch produced products. How could you advertise and display them? Will the selling price be attractive to the **target market**?

Understanding Industrial Practices: Resistant Materials Technology © Nelson Thornes 2004

Working drawings need be full size or to scale. This shows the true proportions of the manufactured parts that are to be made. Working drawings are normally drawn in 3rd angle projection using a **computer-aided design (CAD)** system.

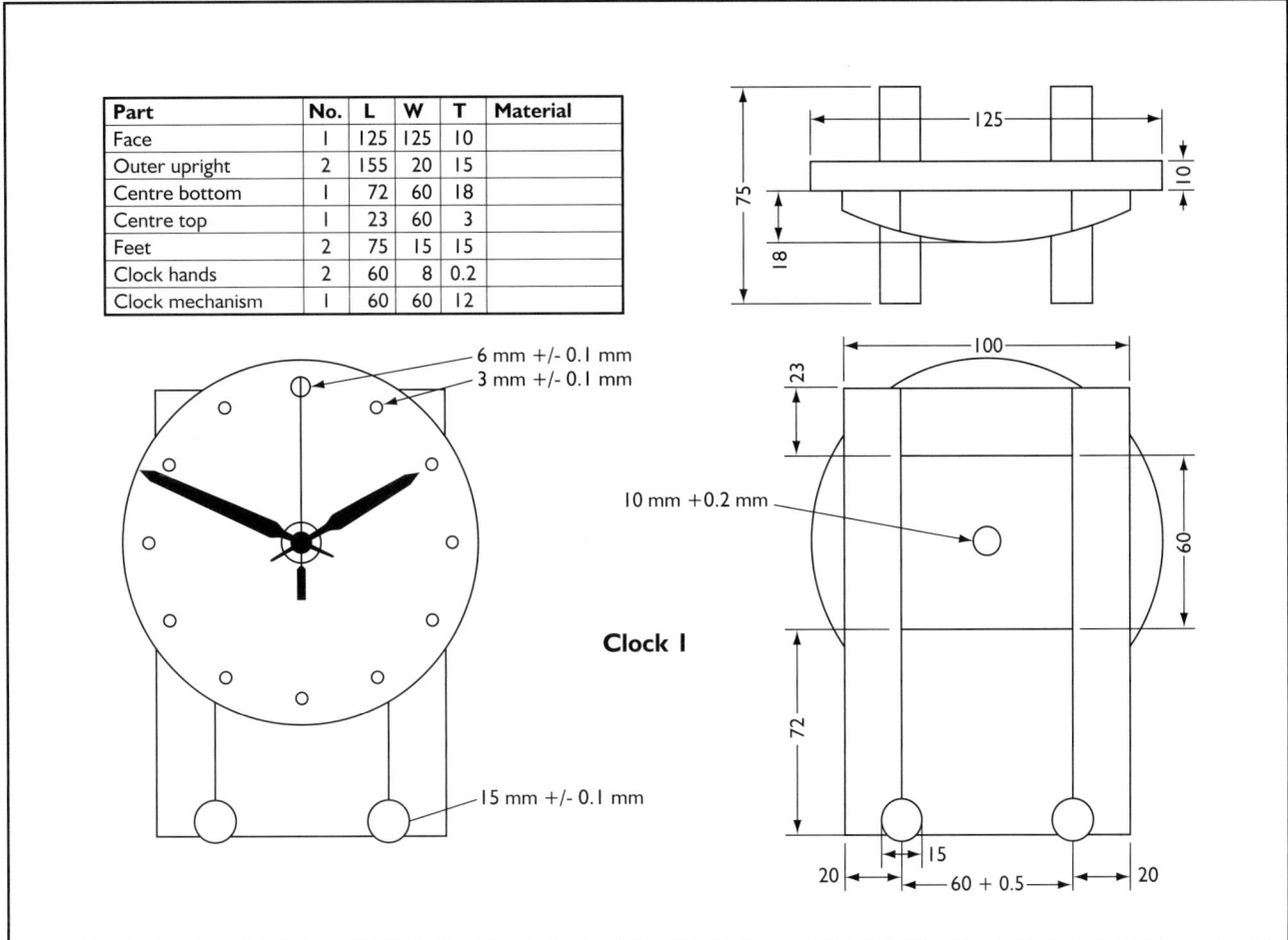

Part	No.	L	W	T	Material
Face	1	125	125	10	
Outer upright	2	155	20	15	
Centre bottom	1	72	60	18	
Centre top	1	23	60	3	
Feet	2	75	15	15	
Clock hands	2	60	8	0.2	
Clock mechanism	1	60	60	12	

Clock 1

1 Above is a 2D CAD drawing of a simple clock. Note that the specific materials have not been given, so you could put in your own. The views shown are from the front, back and above.

   a) Sketch the clock freehand in 3D.

   b) Make a 2D copy of the above drawing to the exact sizes shown on to an A3 sheet using CAD software. Add the major dimensions to the views (height, width and depth).

2 Although most working drawings are in black and white outline (including the one above) try colouring a copy of your CAD views to make the clock appeal to different age ranges or environments. For example, if you colour the clock to look wooden it will look softer than if you use primary colours.

3 Save your original CAD views then modify the shape of the clock to see if it looks better taller or fatter. Try altering the shape of the round face to make it square or star shaped.

Further work

1 List the benefits to **manufacturers** of using CAD.

2 Use CAD to draw to scale a simple product you have already sketched such as a small box, note-holder, small tool or something similar. Remember to include details of dimensions.

Name . . . .

Form/group . . . . **Date** . . . .

Computer numerically controlled (CNC) machines are used in industry to enable the automated production of manufactured products. This is called **computer-aided manufacture (CAM)**.

In the table below is a list of CNC machines used by industry. For each machine, list the manufacturing process it is used for and the type of products that are made. Some rows have already been completed for you.

CNC machine used in industry	Manufacturing process	Type of products produced
CNC lathe	Machining rod, bars and spindles by rotation	Candlesticks, screws, knobs, bushes, spindles
CNC milling machine	Machining slots, holes, grooves and shapes using drill-like or circular blade cutting tools	Specialised metal parts such as disc brakes for bicycles
CNC router	Machining slots, holes, grooves and shapes using drill-like cutting tools	Slot together toys, flat-pack furniture, edge mouldings
CNC drill		
Injection moulder		
Extruder		
Blow moulder		
Vinyl cutter		
Printer	Printing	Templates to support manufacture

Further work

1 List the benefits to **manufacturers** of using CAM.

2 Explain how you could use **computer-aided design (CAD)** and CAM to improve the **quality** of your own work.

3 Make a list of the CNC machines available to you in school and say how they could support your project work.

Name . . . .

Form/group . . . . **Date** . . . .

Computer systems play an important role in product manufacture.

1 Give three examples of the general use of computer systems in manufacturing industry.

2 Give three examples of the use of specialist computer systems in product manufacture.

3 Describe the benefits to **manufacturers** of using computer systems.

4 Using the input-process-output block diagrams shown below explain how you could use a computer system to help you:

   a) collect and analyse data about products

   b) produce **working drawings** ready for manufacture

   c) produce a sample product using a **computer numerically controlled (CNC) machine**.

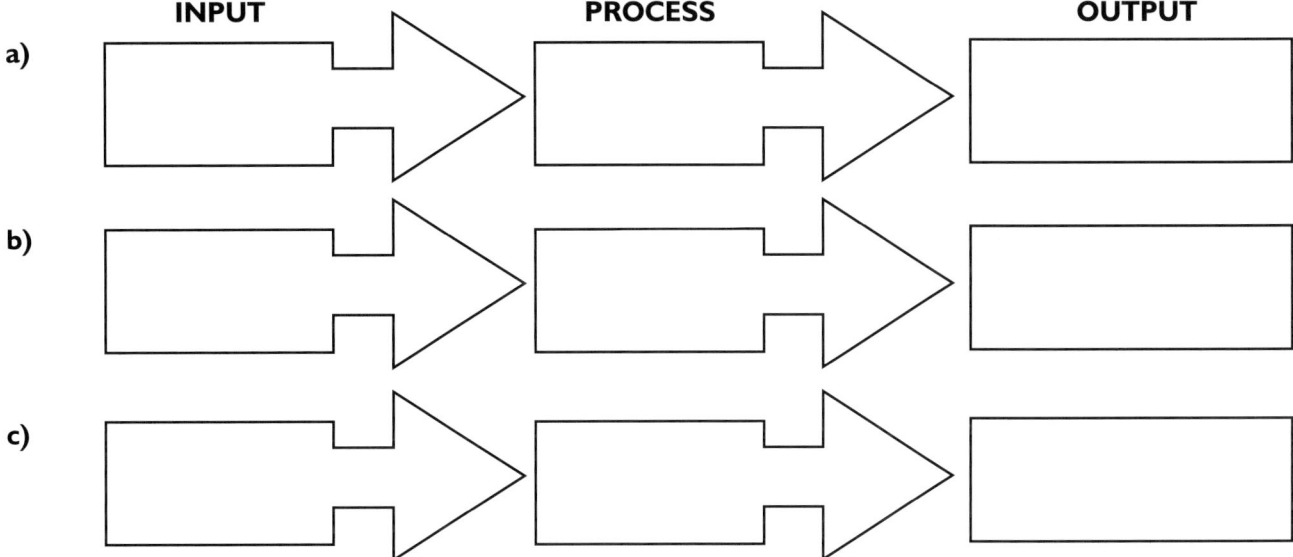

Further work

Collect information about all the computer software and machines available to you in school and list those that may support you in your design and make activities.

Software & CNC machine	Design and make activity

7

construction

Name . . . .

Form/group . . . . **Date** . . . .

I Complete the table below to show the **joining system** that could be used to construct the corner of a wooden toolbox. Assume that glue is also used.

Joining system	Solid wood		Manufactured board		Strength	Difficulty
	Suitability	*Looks*	*Suitability*	*Looks*	(high, mid, low)	(high, mid, low)
Finger or comb joint	good	good	good	good	high	mid
Dovetail joint	good	good	poor	moderate	high	high
Screw joint						
Peg or dowel joint						
Half-lap joint						
Butt joint						
'Bloc-joint' (knock-down fitting)						

2 Complete the table below to show the system that could be used to join the corners of a tubular steel or tubular aluminium frame.

Joining system	Steel		Aluminium		Strength	Looks
	Suitability	*Difficulty*	*Suitability*	*Difficulty*	(high, mid, low)	
Heat weld	good	moderate	good	high	high	good
Screws or bolts						
Rivets						
Adhesive (e.g. Araldite)						

3 Complete the table below to show the system that could be used to join the corners of a plastics storage box. Assume that corner brackets may be used with screws, bolts or rivets.

Joining system	Acrylic		Polythene		Strength	Looks
	Suitability	*Difficulty*	*Suitability*	*Difficulty*	(high, mid, low)	
Screws or bolts	moderate	low	moderate	low	mid	moderate
Rivets						
Solvent adhesive						
Heat weld						

Further work

Sketch the joining systems used for constructing containers and frames in wood, plastics and metal materials.

Understanding Industrial Practices: Resistant Materials Technology © Nelson Thornes 2004

8 continuous production

Name . . . .

Form/group . . . . **Date** . . . .

1 Give three examples of resistant materials products manufactured by **continuous production** techniques. Explain why they are made by this **method of production**.

2 Give three reasons why **one-off** products are more expensive to buy than those made by continuous production.

3 Describe two benefits of using sensors to monitor continuous production machinery.

4 Describe the differences between one-off production, **batch production** and continuous production.

9   control systems

Name . . . .

Form/group . . . . **Date** . . . .

In **control systems**, **flow diagrams** or block diagrams are used to show how inputs, processes and outputs are used to breakdown the **system** into smaller parts. This helps to organise and manage the system in a more efficient way.

1   Draw a systems diagram to show an open loop control system for making identical rectangular, finger jointed wooden boxes.

2   Decide how **feedback** could improve the **quality** of the boxes and draw a closed loop control system showing where this feedback occurs and what checks would be made.

3   Draw a systems diagram to show how you could undertake each of the following activities:

- use a computer controlled cutter (plastics, wood or metal) to produce identical component parts for a small toy with a cam action feature

- manufacture your current resistant materials product with specified checks to improve quality.

Understanding Industrial Practices: Resistant Materials Technology © Nelson Thornes 2004

10   costing

Name . . . .

Form/group . . . . **Date** . . . .

A simple polythene bag 'carry-aid' is shown in the diagram. It is used to reduce hand fatigue when carrying full supermarket polythene bags. It is made from (100 × 200 × 3 mm) bent acrylic using a strip-heater and has a computer-cut vinyl logo stuck to both sides.

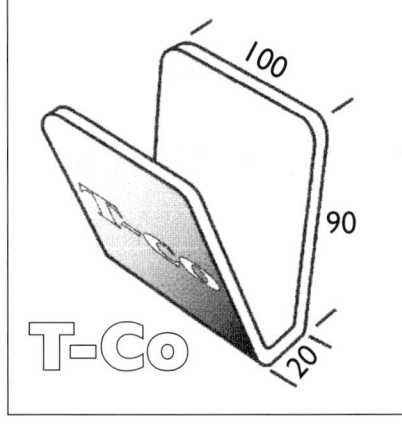

Work out the cost of making a small batch run of 10 carry-aids in the set stages shown below:

a) Work out the Direct Costs for the carry-aid.
Direct Costs = material costs + labour costs

    Material costs for 10 carry-aids
    Acrylic (1 metre × 0.2 metres) @ £10 per square metre   =
    Vinyl logos (1 metre × 0.1 metre) @ £2 per square metre =

    Labour costs
    60 minutes @ £5.40 per hour   =
    Total Direct Costs   =

b) Work out the Overhead Costs for 10 carry-aids.
    Overhead Cost   = 10 % of the labour cost
                                          =

c) Work out the Manufacturing Costs for 10 carry-aids.
    Manufacturing Cost   = Total Direct Costs + Overhead Costs
                                          =

d) Work out the Manufacturing Profit for 10 carry-aids.
    Manufacturing Profit   = 20 % of the Manufacturing Costs
                                        =

e) Work out the Selling Price for 1 carry-aid.
    Selling Price of 10 carry-aids   = Manufacturing Costs + Manufacturing Profit

Further work

1 The selling price of the carry-aid may be too high for it to be given away to supermarket customers. Try to work out how you could reduce the unit cost of the carry-aids if you were to **batch produce** 100 using school facilities. Use the diagram and information to consider:

    a) the type and size of **material**

    b) the manufacturing process.

2 Work out a new selling price to the supermarket for each unit based on your new design. Use the method outlined above. For this exercise, leave the time it takes to make each unit the same as above.

3 If you found it was successful and 100,000 were to be produced how could the unit cost be further reduced? Consider the type of material, the manufacturing process and the use of industrial production methods.

Analyse a simple manufactured product such as a modern wall clock, bottle-opener, scissors or pen.

1   Sketch the product and work out the number of component parts and the processes needed to make it.

2   Draw a **flow diagram** similar to the one below to show the **stages of production** for the product you have chosen.

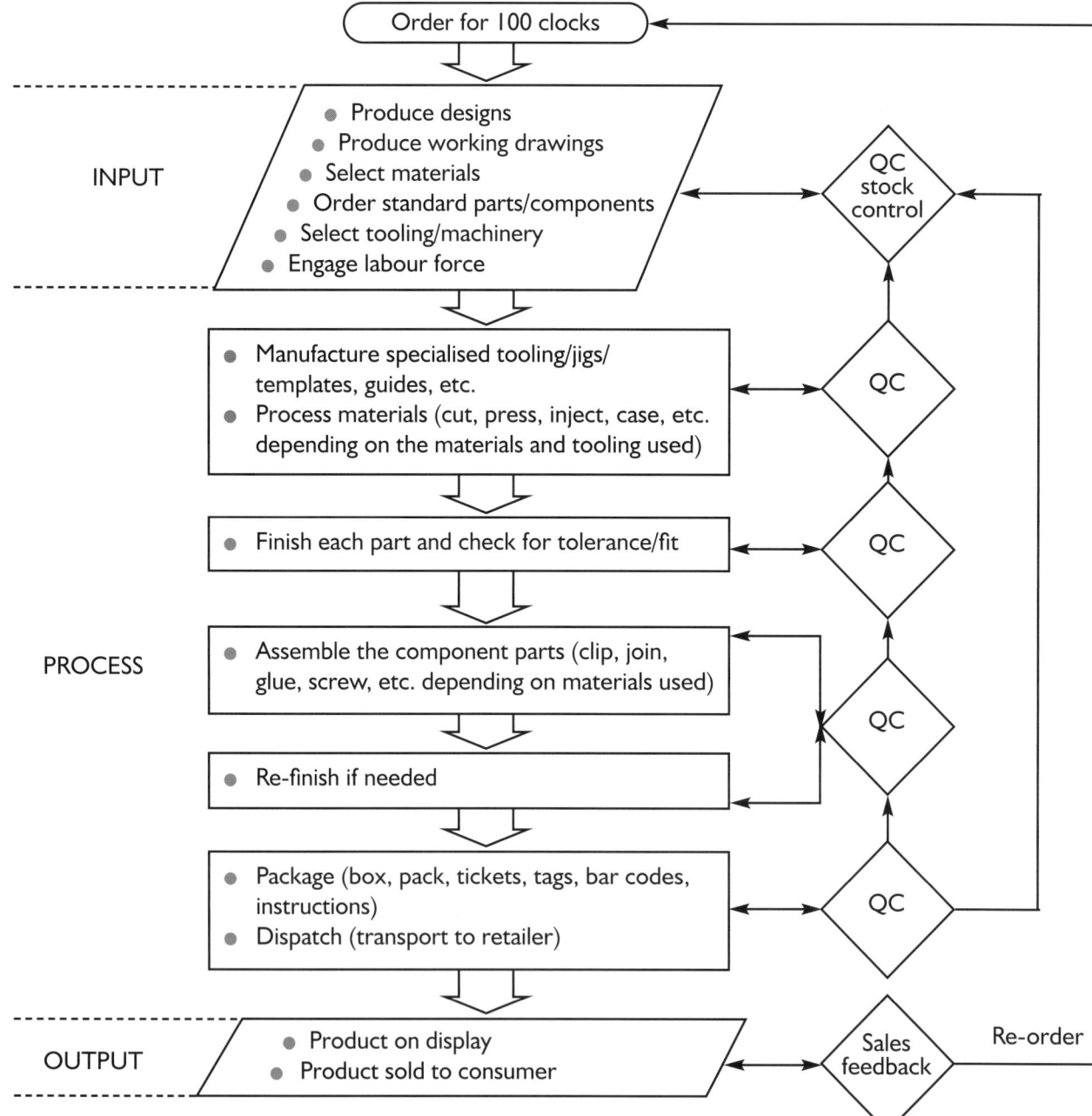

Further work

1   Describe the processes used to make the product.

2   Explain where **critical control points** would occur in your flow diagram. Describe the **quality indicators** you would use to check the **quality** of the product.

Name . . . .

Form/group . . . . **Date** . . . .

You have been commissioned by an Exhibition Centre to design a simple, low-cost, flat-pack or folding seat that would be available for visitors to use free of charge in the centre. The seats are to be manufactured as a small batch of 10. A compact storage system also needs to be designed for the batch of 10 seats. This can be either a 'stand' or a hanging system for the wall.

Write a **design brief** to describe:

- the type of products you will design, such as a compact folding or flat-pack seat and a 'stand' or wall mount hanging system
- the purpose of the products or what the seats are to be used for, how long users are likely to be seated at any one time and how easily the seats need to be erected and put away
- the likely **target market** who would use your products. Include their age group and size range.
- the environment and aspects of **safety**, for example, risk of fire, trapped fingers, collapsing
- a time limit for generating the ideas.

Further work

Research resistant materials products that fulfil a need for teenagers who work, rest, play and sleep in the same room.

Write a design brief to describe:

- the type of product you will design to fulfil a need
- the target market group
- the **end-use** of the product.

13 writing a design specification

You have been commissioned by a specialist lighting retail store to design a desk light for a young person's study area. It should be appealing, low cost and sell well to a teenage market.

Undertake some initial research based on your analysis of the **design brief** for a young person's desk light. Write a **design specification** for the light using the following questions to guide your thinking:

- What is the function of the product?

- What are the quality requirements of the **client**, **manufacturer** and the **target market group** (**consumers** and users), relating to the **performance requirements** of the product, **materials** and components? (Include attributes, **materials**, **finish**, **size**, weight, maintenance, etc.)

- What should the product look like in order to be attractive to the target market group?

- How much will consumers be prepared to pay? What is the price range of similar products available in the market? What is an acceptable cost limit for manufacture?

- What **safety** checks need to be considered for the product? (Size, weight, **colour**, finish, sharp edges, material type, electrical, etc.)

- What social, legal or **environmental issues** are relevant and need specifying?

- How many products are needed (**one-off**, **batch**, **high volume**)?

- How much time is there to design and make the product?

Understanding Industrial Practices: Resistant Materials Technology © Nelson Thornes 2004

14   using a design theme

A **design theme** can help you to focus your research and develop **style** and **colour** ideas for your product. It can sometimes be developed from the **design brief** or be built around the needs or lifestyle of the **target market group**. For example, you might be asked to design a new sit-on toy for children using a 'Jungle' theme. This could focus on animal and plant jungle styles and colours.

1   Develop a theme based on a design brief.

Design brief
Design and make a sit-on toy for the under 5s age group. It needs to be suitable for use both indoors and out. You could develop a theme for this in the following way:

- Read the design brief carefully and pick out key words to help you develop the theme.
- Use the theme to help you collect images and colour ideas.
- Put together a **mood board** to use as a starting point for generating ideas.

2   Develop a theme based on the needs of a target market group and/or a specific retail outlet.

Needs of the target market group
A consumer/user profile can provide details about the likes, dislikes and lifestyles of **consumers** in the target market group. These consumers often buy products in specific retail outlets. The consumer/user profile helps to build a picture of the kind of product that the target market group wants or will buy. This enables **manufacturers** to target the right retail outlet with the right product at the right price.

Write a consumer/user profile based on the needs of young children who are learning to count. Take into account the need for a stimulating play and learn experience. The consumer/user survey could include information about:

- where parents shop when looking for 'educational' toys
- gender
- age group
- **brand** loyalty
- television programmes children watch
- how children play safely with toys.

Use the information you collect to write a design brief based on the theme 'Zoo'.

Further work

Designers sometimes base ideas on themes such as 'The Egyptians', 'Space' or 'Nature'. They also make use of things they can easily collect, see or experience and may use social or **environmental issues** to influence design themes.

Use the influence of one of the following themes to develop a design brief for a product of your choice:

- Products from junk
- Green peace
- Animal magic.

When you analyse any manufactured product you should aim to identify the **properties** of the **materials**, the **construction** techniques, the **finish** and maintenance requirements.

- Choose a simple product such as a biro-pen, paper punch, 13 amp plug, toothbrush, glue stick or small tool.
- Use the table below to help you analyse the materials and construction techniques used in the product.
- Use the table on worksheet 16 to record your findings.

Task	How to carry out your task
Identify the materials	- Examine and list all the component parts of the product (such as body, hinge or fastener, inserts) - If the part is made from wood, is it hardwood or softwood? Why is this type of wood used? - If it is plastic what type is it and why is it used? - What are the hinges or **fasteners** made of and why?
Identify the construction techniques	- What construction technique is used to make each component part? (Machined, cast, formed, moulded, etc.) - How is each component part fixed to the next? (Glue, screw, clip, etc.) - How does the construction affect the product styling?
Investigate the properties and working characteristics of the materials	- Examine the materials used in the product. Are they smooth or shiny, tough or scratch resistant, for example? - Compare what you find out about the properties of the materials used in the product with their expected properties. - Explain why the properties of the materials make them suitable for the product. - Why do different **manufacturers** use different materials for the same product (e.g. kettles, seats, pens, toys)?
Investigate the finish	- What type of finish was used on the product? - How was this finish applied? - What is the purpose of the finish?
Describe the product's maintenance requirements	- If the product is used regularly how can it be kept in good condition?

Further work

1 Draw up a table for three products made from wood, three products made from metal and three products made from plastic. Identify the type of wood, metal or plastic used in the products. List the properties of each material.

2 List and compare a range of suitable finishes for wooden and metal garden furniture.

Name . . . .

Form/group . . . . **Date** . . . .

Use the table below to record your findings when you analyse product materials, **construction** techniques, **finish** and maintenance requirements (see worksheet 15 for guidance).

Task	My findings
Product type	
Identify the **materials**	
Identify the construction techniques	
Investigate the **properties** and working characteristics of the materials	
Investigate the finish	
Describe the product's maintenance requirements	

Further work

Using catalogues or actual products, explore a range of similar types of product that have a small number of component parts. For example, you could investigate garden tools.

If you do this as a team you can each explore one product type and share the results with other members of your team.

- Sketch each product.
- Compare the materials, construction and finish used in each product.
- For each product, list the material properties required by the user.
- Compare the price of each product to determine the best value for money.

17 ecolabelling

Name . . . .

Form/group. . . . . **Date** . . . .

The Product Life Cycle monitors the environmental impact of a manufactured product from cradle (raw **materials**) to grave (disposal). It assesses which part of the product's life cycle causes the most harm through the production of waste, pollution, the consumption of **energy** and natural resources and its effect on the planet's ecosystems.

Use the Product Life Cycle table below to assess the environmental damage that may be caused during the life cycle of a 12-pack of blow-moulded plastics drink bottles.

Environmental damage	Product Life Cycle				
	Raw materials	Manufacture	Distribution and packaging	Use	Disposal
Waste					
Soil pollution					
Water pollution					
Air pollution					
Noise					
Energy consumption					
Demand on natural resources					
Effect on ecosystems					

1 Identify the stage(s) in the life cycle table where you think the bottle pack will have the greatest environmental impact. Mark a tick where this happens.

For example, all plastic products are made from oil, which is a non-renewable resource, so you could put a tick against 'Demand on natural resources'. When burned or disposed of as waste, plastics also have a considerable effect on ecosystems. Put different coloured ticks in the chart where you think other kinds of environmental damage occur.

2 For each environmental damage you have identified, explain what it is and why it happens.

3 Try to work out how the environmental damage you have identified can be reduced or avoided.

Understanding Industrial Practices: Resistant Materials Technology © Nelson Thornes 2004

18   feedback

Name . . . .

Form/group . . . . **Date** . . . .

If a product is selling well, **retailers** can **feedback** sales information electronically via an **Electronic Point of Sale (EPOS)** till. Production can then be increased to meet demand. This kind of sales **system** can be shown using an input/process/output block diagram or **flow diagram**.

1   Draw an input/process/output diagram with feedback to illustrate an effective sales system that includes feedback. You need to consider manufacture, delivery, stock control and sales.

2   Explain the difference between an open loop and closed loop system.

3   Draw a flow diagram to show the use of feedback in a system used to manufacture a simple wooden pencil box. You will need to include: ordering and cutting materials, the **construction** of **joints**/fittings, **assembly** and **finishing**. Show where feedback occurs in the system and describe the **quality indicators** required to provide feedback to the system.

19   flow diagram

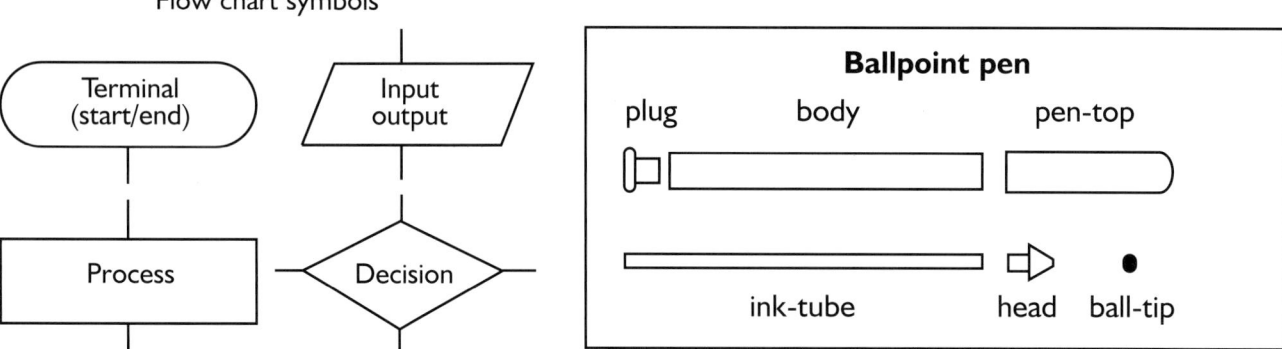

The main manufacturing and **assembly** processes for a simple stick ballpoint pen are given below. However they are in the wrong order. Using the appropriate standard symbols shown above, draw a **flow diagram** to show the manufacturing and assembly processes in the correct order.

- press the ball-tip into ballpoint head

- mould the pen-top

- fill with ink

- assemble ink-tube and head to body

- mould the body end plug

- assemble head and ink-tube

- insert body end plug

- assemble body and pen-top

- make the ballpoint head

- mould the body

Further work

1 State what **quality** checks would be made during the production of the ballpoint pen, such as checking for fit or leaks. What action would be taken if the ballpoint pen failed a quality check?

2 In your flow diagram of the manufacturing and assembly processes for the ballpoint pen show where quality checks would be made.

3 Draw a flow diagram to show the processing and assembly stages in the correct order for your own product. Introduce quality checks and draw **feedback** loops.

Understanding Industrial Practices: Resistant Materials Technology © Nelson Thornes 2004

20 hazards and hazard analysis

Name . . . .

Form/group. . . . . **Date** . . . .

The table below shows the **stages of production** for a manufactured product.

	Potential hazards		
Stages of production	Workforce	User	Environment
Preparation			
Processing			
Assembly			
Finishing			
Use			
Disposal			

1 Using the table, identify the potential **hazards** in the preparation, processing, **assembly**, **finishing**, use and disposal of your own coursework project. Decide who or what is at risk, i.e. the workforce, the user, the environment.

2 Explain how you could make changes to the product and its manufacture in order to make the product safer.

Remember to check that you have included **safety** criteria in your **design** and **manufacturing specifications**.

21 high volume production

Stages of production	One-off product	High volume product
Preparation		
Processing		
Assembly		
Finishing		

a) Choose a **one-off** product you have already designed and made.

b) Draw up a table similar to the one above to show the four **stages of production** for your one-off product.

c) In the right hand column, identify any changes you would need to make to your one-off product if it was to be made in **high volume**. You will need to think about the following:

  ● How you could simplify the design to make it easier to manufacture.

  ● How you could use standard size **construction** material or **standard components** such as screws, knobs, handles, wheels, hinges and brackets, etc.

  ● If you could buy the **materials** and components in bulk to reduce manufacturing costs.

  ● How you could use **computer-aided design (CAD)** to improve the design and manufacture of your product.

  ● How you could use **jigs** or **templates** to speed up production.

  ● How you could use **quality control** and **feedback** to make identical products.

Further work

1 Explain the difference between one-off, **batch** and high volume production.

2 Give an example of one product that could be made by each **method of production**.

3 Compare the manufacture, cost and value for money of a jewellery box made by hand with one that is produced in high volume. What design features would make the one-off box value for money?

Understanding Industrial Practices: Resistant Materials Technology © Nelson Thornes 2004

22 industrial practices

Name . . . .

Form/group. . . . . **Date** . . . .

1   You have been asked by a **client** to design and manufacture a small portable picnic table that could be carried flat in the boot of a car and sold in camping equipment shops. The first production run would be for 100 tables.

● List the main responsibilities of the client for this product.

● What must the **designer** agree to, find out or consider for the product to be successful?

● What sort of **materials** and manufacturing processes could be used to manufacture 100 tables?

● What is the most important need of the product **manufacturer** and how could it be met when designing the picnic table?

● What are the **user** requirements for a product of this nature?

Further work

Identify six industrial designing activities and six industrial manufacturing activities that you could use in your own coursework project.

Designing activities

Manufacturing activities

23 ICT, CAD and CAM

Name . . . .

Form/group . . . . **Date** . . . .

Use **Information and Communication Technology (ICT)**, **computer-aided design (CAD)** and **computer-aided manufacture (CAM)** where appropriate in your design work, but do not use them where a quicker or better method is possible. Sometimes it is more appropriate to draw or sketch by hand, especially when you are developing first ideas. Using ICT or CAD in your project may depend on what is available to you and how much time you have.

1 If you have access to graphics, word processing or spreadsheet software you can:

- present annotated research using scanned images, digital photos, CD-ROMs or the Internet

- produce documents, create **materials** or component comparison tables and draw **flow diagrams** to support your planning, ideas and production

- produce materials lists and **costing** tables for your product that can be quickly updated or changed

- make use of annotated digital images of the production process as your product develops.

2 If you have access to CAD software you can:

- develop your ideas on screen after the initial sketching of ideas is complete

- develop your own library of design ideas

- render your drawings on screen for quick visual **colour** impressions

- rotate your drawings as 3D models on screen before production

- make accurate, scaled **working drawings** to illustrate dimensioned component parts

- create a **layout plan** to make economical use of cut sheet material

- produce printed **templates** to mark out component parts accurately ready for cutting.

3 If you have access to CAM equipment you can use CAD to:

- produce working drawings of component parts ready to import into a **computer numerically controlled (CNC)** machine such as a vinyl cutter, router, laser cutter, mill or lathe.

Remember that using computers should speed up the process of design and manufacture so use them for the things they are good at.

If you have access to a computer at home you may be able to make good use of it for your school projects.

Understanding Industrial Practices: Resistant Materials Technology © Nelson Thornes 2004

There are many types of **jigs** and **templates** that you can design, make and use to help you with your project. If any process is repeated more than once it is likely that a jig or template will be useful. It is important that your jigs are made accurately or there is little point in using them.

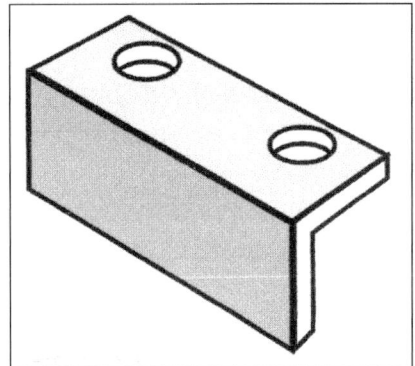

A simple steel jig can be used for making dowel or peg joints in wooden frames.

1 Design a jig that would help you to make accurate comb joints repeatedly.

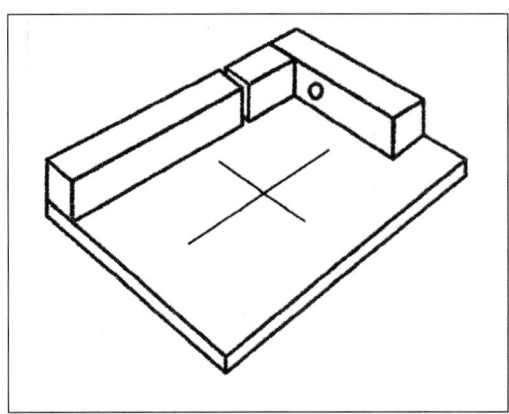

The jig shown to the left is made of MDF and could be used to cut material to the same length. The same jig could be used as a guide to hold material in the same place when drilling holes using a pillar drill.

2 Design a jig that would help you to cut 16 dowel pegs of 8 mm diameter to lengths of 30 mm for a frame joint. You will need to consider how the dowels will be held when they are being cut.

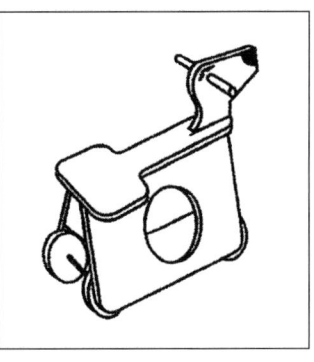

3 Design a jig that could be used to locate the drill holes for the wheel axles of a child's plywood sit-on trike such as the one shown. The holes must be an exact distance from the bottom edge and an equal distance from the front and back edges. You will need to consider and describe how the jig will be held in position.

4 Design and make a simple jig for your own product.

Templates could have been used to produce the shape of the head and the seat of the trike above.

5 Design a template that could be used to draw the head shape and locate the eye positions for a small 'animal-style' pull-along toy suitable for a young child.

6 Design and make templates to help you accurately mark out the component parts of your own project.

Just in time (JIT) planning makes use of **Gantt charts**. The Gantt chart below gives a manufacturing time plan for a simple sit-on trike made from painted plywood. It could easily be made by hand in a school workshop or in low volume using a **computer numerically controlled (CNC)** router.

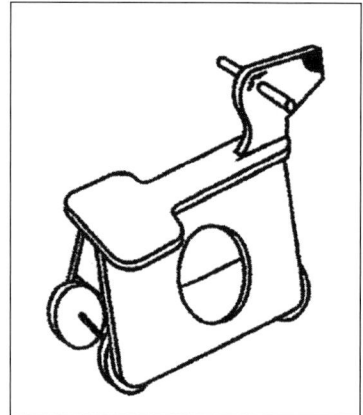

The estimated manufacturing time is marked with an 'X' in the box. If something unforeseen occurs, such as illness or the late arrival of **materials** or components, the manufacturing time may run beyond the deadline. It is good practice to use a different symbol on the Gantt chart to show the actual time of manufacture as work progresses.

1   Produce a Gantt chart similar to the one below to show the deadlines for the manufacture of your product. In the task column, list the processes required to manufacture the product. Mark in the estimated time for manufacture.

	Weeks									
Task	1	2	3	4	5	6	7	8	9	10
Buy the wood, wheels and wheel fasteners, screws and paint	X									
Produce layout plans from working drawings for ply sheet	X									
Produce templates for marking out the sheet ply pieces and mark out		X								
Cut out the sheet ply pieces and drill the screw and axle holes		X	X	X						
Make the front hand grip and fit to the head					X					
Paint the sheet ply pieces before final assembly					X	X	X			
Make and assemble the steel axles and fit the wheels							X	X		
Screw the sheet ply pieces together								X	X	
Test the whole assembly for strength and make a final quality check									X	

Gantt chart for manufacturing a low volume sit-on toy to be made in painted plywood using CNC machines

2   List all the materials, components and the equipment required to make your product.

3   Identify when you will need to have materials and components available for manufacture.

4   Check the availability of materials/components and, if necessary, plan the ordering of them in advance.

5   Explain when and how you would check the **quality** of the materials and **construction**.

6   Mark on the Gantt chart the actual time of manufacture as your work progresses.

26   manufacturing specification (1)

The analysis of a successful manufactured product can help your understanding of the importance of a **manufacturing specification**.

1   Examine an inexpensive or a high cost manufactured product, such as a piece of garden equipment. Ask the following questions:

- What makes the product successful? Consider the design features, **style**, **colour** range, cost of manufacture and **brand** image.

- How does it meet the **quality** and cost needs of its **target market group**? Is it **fit for its purpose**, fault-free, durable and easy to use and how have costs been kept competitive?

- How is the product manufactured and assembled? Is it simple or complex with many separate parts and fastenings?

- What types of **materials** and components have been used? Is the product easy to maintain?

- How well is the product made? How well do the components function?

- How has the product been made safe to use? Is it made from non-toxic materials? How have sharp edges been reduced? How are the fastenings secured? How can the product be easily recycled or made safe for the environment after disposal?

2   Record your answers then produce a manufacturing specification for the product.

Include a precise written description and **working drawings** of the final product. Working drawings should show detailed views of the component parts that include dimensions and critical **tolerance** limits. For this exercise the working drawings may be hand sketched for speed. Remember to include materials and components lists and details of colour and surface **finish** for each part. Also include product performance criteria such as how long the product is expected to last in normal use.

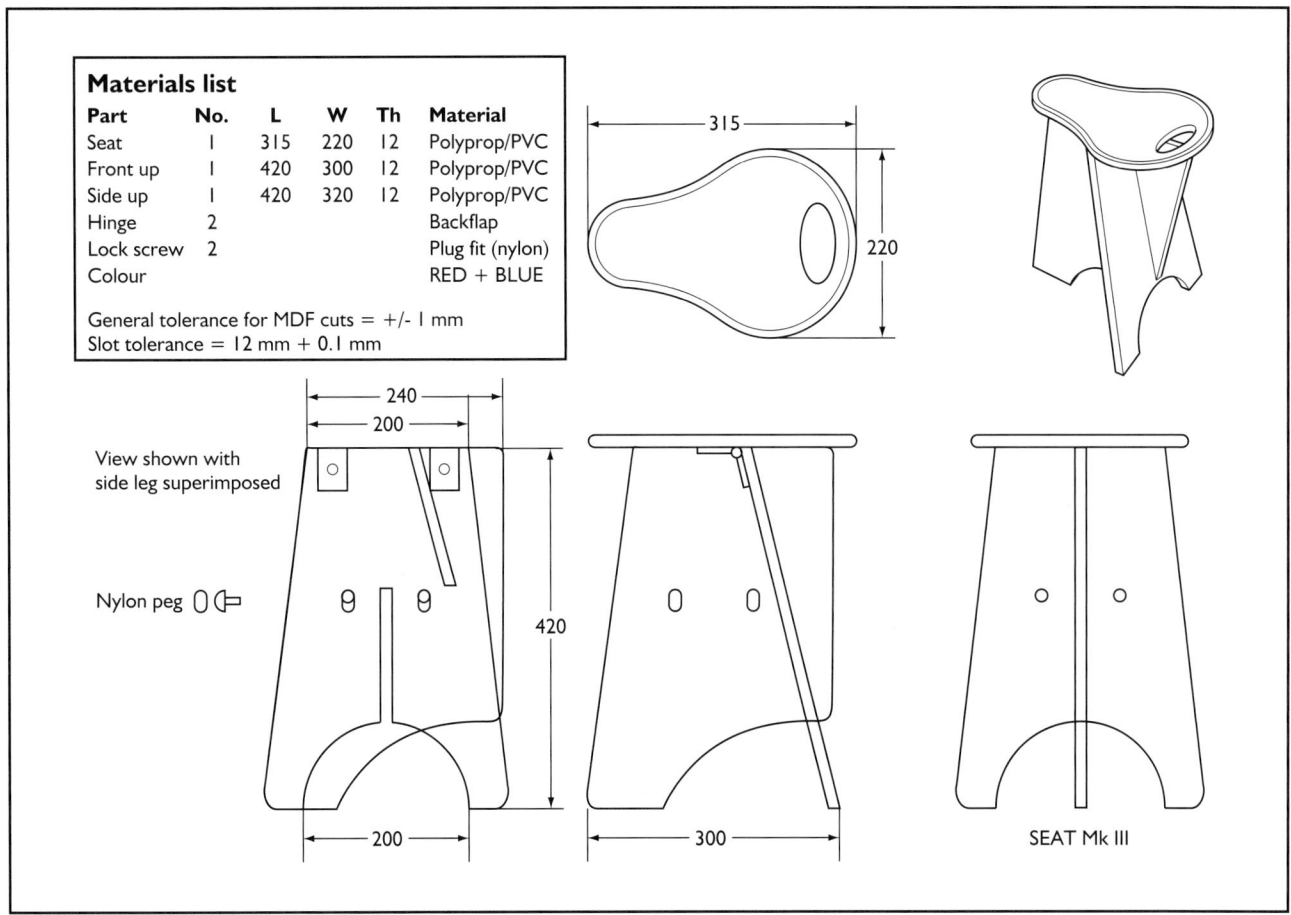

Materials list

Part	No.	L	W	Th	Material
Seat	1	315	220	12	Polyprop/PVC
Front up	1	420	300	12	Polyprop/PVC
Side up	1	420	320	12	Polyprop/PVC
Hinge	2				Backflap
Lock screw	2				Plug fit (nylon)
Colour					RED + BLUE

General tolerance for MDF cuts = +/- 1 mm
Slot tolerance = 12 mm + 0.1 mm

View shown with side leg superimposed

Nylon peg

SEAT Mk III

The view above shows a typical **working drawing** that supports the **manufacturing specification**.

1 Produce a written description of your current product that includes the following information:
- A general description of the product and its features.
- Product performance criteria (such as how much weight it may have to safely support. This should help with **testing** and **quality control**).

2 Produce a working drawing of your current product, similar to the one above that provides the following information:
- Drawings to show detailed dimensioned views of the component parts.
- All major dimensions and critical **tolerance** limits.
- **Material** and components lists including details of bought-in component parts, fastenings or fixtures.
- Details of **colour**, **finish** or surface texture for each part.

Further work

1 Ask someone else to explain how to make your product using only the information you have given in the manufacturing specification. You are checking to see if enough information has been given.

2 Check the **quality** of your product against the details, dimensions and critical tolerances given in your manufacturing specification. This will help to ensure that you produce a high quality product.

Wood

1 You have been asked to design a low cost set of wooden kitchen furniture, which includes a table and four chairs.

- Name the specific type of wood that is suitable for the table and chairs and give reasons for your choice.

- List the **joints** that would be suitable for the chairs if they were made as a batch of 200.

- Sketch one joint suitable for the chair and specify the sizes of the component parts.

- The table and chairs need to be designed as a very expensive, high **quality** set. Name two specific types of wood you would use and give reasons why they are suitable.

- Explain how the **construction** of the high cost set of table and four chairs would differ from the lower cost set.

2 MDF is used by industry to make kitchen cupboards.

- State the characteristics of MDF that make it suitable for kitchen cupboards.

- Describe how MDF is finished commercially to make it suitable for kitchen cupboard doors.

- Describe the **safety** precautions you would need to take when cutting MDF with a power jig saw in a school workshop.

3 Plywood is light and very strong for its weight. However, it is more expensive than MDF and can splinter on its edges.

- Give one reason why plywood always has an uneven number of layers.

- Using notes and sketches, describe two ways that a manufacturer could fix thin plywood to the back of a cabinet.

Plastics

1 Plastic chairs or seats are used in many classrooms and offices.

- Name a specific plastic that is suitable for making a batch of 50 'tulip shaped' office chairs. Name the method of construction you would use.

- A manufacturer wants to design and make 100,000 classroom chairs in plastic. Name a suitable plastic and describe the process suitable for manufacturing the chairs.

2 A plastics bearing is needed for the 8 mm diameter axle of a child's wooden sit-on pedal car.

- Name the specific plastic material suitable for the bearing if it was made in low volume using school workshop techniques.

- Sketch the shape of a suitable bearing and illustrate how it could be held to the car frame.

3 A very strong, rigid plastic is used for making CDs.

- Name the specific plastic material that is suitable for the CDs. What different plastic would be suitable for the low cost, clear plastic 'jewel' case the CD is kept in?

- Name the specific type of softer plastic used to make the 'cake' style cases that hold up to 100 CDs.

Metal

1 You have been asked to design and make a metal toolbox for storing spanners. The size is 400 × 200 × 150 mm.

- Name two specific metals that would be suitable.

- Describe two techniques you could use to make the box corners in one of the metals you have chosen.

- Name two industrial techniques that could be used to manufacture 1,000 of the metal boxes.

- Explain how you would finish your chosen metals.

2 Name the specific metal used to make most spanners. Describe how they are made and what **finish** is used.

Disposal of materials

1 In the environment ('env') columns of the table below number the materials according to how much damage they cause to the environment (1 = least damaging, 12 = most damaging).

Material	Order		Material	Order		Material	Order		Material	Order	
	env	dcp		env	dcp		env	dcp		env	dcp
Oak			Bio-degradable polythene			Lead			Acrylic		
Steel			Pine			Plywood			Aluminium		
Mahogany			UV resistant PVC			MDF			Tin-plate		

2 In the decompose ('dcp') columns of the table above number the materials according to how quickly they decompose (1 = very quickly, 12 = very slowly). Pine is the first to decompose so this should be numbered 1.

3 Explain how the impact on the environment can be reduced when products have finished their useful life.

4 Describe what you or your family do when a large resistant materials product has past its useful life and cannot be sold on (e.g. a rusting bicycle). What happens to it next?

5 Name eight materials or products that can be made from recycled waste.

6 List three different specific materials from which drinks containers are made. Compare the advantages and disadvantages for the **manufacturer**, users and the environment of the use of each material on your list. Suggest how each material could be recycled or reused.

Further work

Make a list of questions that will help you decide the best materials from which to make your next end product.

Name . . . .

Form/group . . . . **Date** . . . .

1 Choose a simple manufactured product such as a clock, pendant, small box or simple tool that could have been made in a school workshop by a **one-off production** method.

My chosen product is: ....

2 List the processing and **assembly** stages used to make the one-off product.

3 Describe three ways in which the **method of production** could be improved, or simplified, to make it more suitable for manufacturing in **high volume**. You need to think about how to:

- simplify the design
- use **standard components** or fastenings
- use alternative **materials** that may cost less
- reduce the amount of material or the decoration used in the product
- control the **quality** to make each part accurately
- use **templates** and/or **jigs** or **computer-aided manufacture (CAM)**
- change the manufacturing processes and simplify assembly
- manufacture the product more efficiently using a team of 3 to 6 people.

4 Give three reasons why it costs less to manufacture a product in high volume.

Name . . . .

Form/group . . . . **Date** . . . .

You have been asked to design a range of candleholders suitable for a festive occasion.

1   Sketch three annotated ideas for simple candleholders that would be suitable for your chosen festival.

2   Choose a suitable **method of production** for your candleholders. They must be quick and easy to manufacture. You need to think about:

- the **scale of production**, i.e. how many are to be produced

- how long you have to make them in order for them to be ready in time for the festival

- the **materials** and components available to you

- how your products might be decorated, for example, they may be painted or have added decoration such as tinsel, holly, flowers, leaves or beads

- the techniques you can use and how skilled you are

- how and where in the manufacturing process you can use **quality control** checks

- **safety** in use

- how the candleholders could be packaged for delivery.

3   Explain why your chosen method of production is suitable for your candleholders.

Understanding Industrial Practices: Resistant Materials Technology © Nelson Thornes 2004

32   modelling

Your task is to design a wooden garden seat or table for use at a home barbeque and to model it in 2D. Then model the product in 3D using readily available inexpensive materials.

a) Model your garden seat or table in 2D, using drawings generated by hand or computer. This will enable you to test your ideas and work out what you want to do, so you avoid mistakes later on.

You can:

- sketch several different first ideas and show them to your **target market group**

- experiment with **colour** using **computer-aided design (CAD)** or photocopy outline drawings of your final idea and colour these in different ways

- present your final ideas on a design sheet to get **feedback** from teachers, friends and family

- use CAD to present scale drawings of your final idea.

b) Model your garden seat or table to scale in 3D using readily available materials.

You can:

- try out your ideas for any folding mechanisms in card

- design the seat or table from standard sections of wood to reduce the cost of production. Model it using lollipop sticks or similar material.

- make a full size **prototype** joint in low cost materials to test it for strength and for the time it takes to manufacture

- use a digital photograph of your scale model and superimpose it on to a photo of its intended environment to test how the seat looks.

c) Draw a **flow diagram** to show the order of manufacture and **assembly** for your garden seat. Include **critical control points** to show where you would check for accuracy and **quality**.

Further work

1   Model ideas for one of the following: a mechanical action toy (cam or linked lever), a computer workstation, a folding or flat-pack seat or an iron.

2   Explain why **manufacturers** model or prototype products before manufacture.

Analysing a commercially manufactured product can help you understand how other **designers** solve design problems. Choose a product that could be used in a kitchen such as a pepper grinder, toaster, bread bin or kettle. In most cases you will only be looking at the container, unless the product is easily taken apart like a pepper grinder.

1 Sketch the front and side views of the product.

2 Describe the product using the following headings:

Product description

Product cost

Scale of production

Target market group

Materials type

Material **properties**

Construction system

Product maintenance

3 Using a scoring table similar to the one below, evaluate the **quality** and value for money of the product. Score: 5 = excellent, 4 = good, 3 = satisfactory, 2 = below average, 1 = poor

The product	Score	Comment
How good it looks		
Materials and components used		
How well it performs its function		
How well it meets user needs		
How environmentally friendly it is		
How well it is made		
Accuracy of construction		
How well it is finished		
Suitability of fixings/lid fit, etc.		
Value for money		

34 product analysis (2)

Analysing a commercially manufactured product can help you understand how other **designers** solve design problems. Choose a product that could be used in a kitchen such as a pepper grinder, toaster, bread bin or kettle.

Using the key words below to help you, describe the product's features.

a) List and describe three special features about the product.

b List and describe two good points about the product.

c) List and describe two bad points about the product.

d) Using the information you have learned, explain why the **target market** user might buy the product.

KEY WORDS

Function
How well does the product work? How well does it perform its function?

Materials content
- Hardwood, softwood (such as oak, pine).
- Plastics (such as acrylic, polypropylene).
- Metal (such as steel, aluminium).

Materials finish
- Polished, coated, plated (such as wax, cellulose sealer, paint, chrome-plated).

Maintenance
- Cleaning.
- Filling (salt, coffee, etc.)
- Servicing.

Fastenings/fixings/joints
- Permanent: glue, weld, rivet.
- Non-permanent: hinges, screws, pins, clips, locks, use of brackets.

Properties
- Soft, hard, rough, smooth, rigid, dull, shiny, fire-resistant, heavy, lightweight, shiny, flexible, strong, waterproof.

Colour
- Number of colours used in the product.
- Primary, secondary, pastel, bright, neutral.

Style/shape/proportion
- Angular, soft, modern, traditional.
- Overall product size and shape.

Anthropometrics/ergonomics
- The relationship between the shape/size of the user and the product.
- How it suits/fits the user.

Target market
Potential consumers or users of the designed product.

Further work

Using magazines and catalogues choose a resistant materials product suitable for one sporting activity.

- Investigate the **properties** of the **materials** used in the product.
- Investigate the **construction**. Are the fastenings, fixings or joints permanent or non-permanent? How are the fastenings, fixings or joints made?
- Research the needs of people undertaking the sporting activity.
- Draw up a table to show why the materials and **construction** used make the product suitable for the activity.

35 product data management

Name . . . .

Form/group . . . . **Date** . . . .

1   Explain three key features of a **product data management (PDM)** system.

2   Give four benefits of using a PDM system:

- for designing

- for manufacturing

3   The use of PDM systems enables **manufacturers** to make their products on a global scale.

   a) Describe the effect that **global manufacturing** may have on levels of employment in different countries.

   b) Explain what **designers** and manufacturers in 'developed' countries can do to compete with so called 'low wage' economies.

Understanding Industrial Practices: Resistant Materials Technology © Nelson Thornes 2004

36 production line simulation

Name . . . .

Form/group . . . . **Date** . . . .

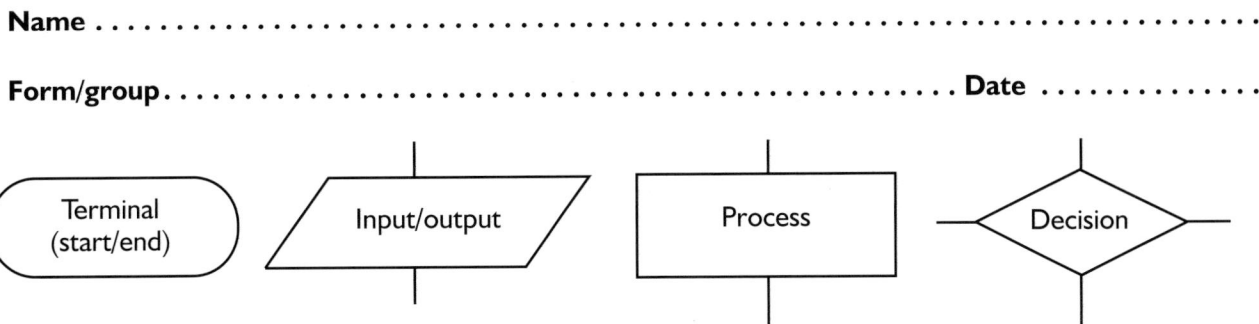

Try simulating a **production line** to manufacture a simple product in **high volume**. You will need to work in a team of 3 to 4 people.

1 Your task is to design and manufacture a simple decorated paper or card container for a single CD. The container should be capable of standing upright on its own for display purposes.

2 Give your team a time limit during which they need to make as many high **quality** CD containers as possible. If you allow yourselves one hour for manufacture do not include the design of the product in this time.

3 Work as a team to plan your production:

- Decide the shape, **size** and decoration of your CD container. Limit the number of **colours** to be used.

- List the different processes, **materials** and equipment needed for manufacture.

- Make a paper **template** for marking and cutting out the CD containers.

- Using the standard symbols above, draw a **flow diagram** showing the manufacturing and **assembly** processes you will use. Show where you will check the quality of your product.

- Work out how long each process will take. Share out the work with your team.

- Manufacture the CD containers with each team member doing one task. Check for quality as you go.

- Run a final quality check and count how many high quality products you have made in the time available. Are all the products made to the same quality?

- Evaluate how well each team member worked. Did anyone have too much/not enough to do?

- If you were to repeat the production line simulation exercise, explain what changes you would make to improve the production system.

Further work

1 Describe the difference between a production line used for high volume manufacture and one used for **batch production**.

2 See worksheet 3: batch production for other ideas on how to set up a design and production team.

Name . . . .

Form/group. . . . . **Date** . . . .

Understanding different types of **properties** will help you choose appropriate **materials** for your designed products.

1   Decide which of the following material properties are **aesthetic** and which are **functional**:

- Shiny
- Warm to touch
- Stain-resistant

- Brittle
- Slippery
- Soft

- Flexible
- Hard
- Strong

- Water-resistant
- Dull
- Durable.

Write the properties in the appropriate column below.

Aesthetic properties	Functional properties

2   You have been commissioned to design a portable, self supporting roadside warning sign for men at work. It must be stable in light to moderate winds and must fold for easy transport. The warning sign itself could be made of flexible material and be held on a frame.

- List each component part of the warning sign. Include the legs/stand, hinges, sign support frame and any other components.

- Write a materials specification for each component part of the product, listing the aesthetic and functional properties of the materials that would make them suitable for the warning sign.

- Explain which of these properties are essential, and which would be useful, for your warning sign.

Understanding Industrial Practices: Resistant Materials Technology © Nelson Thornes 2004

38   prototyping

Making a **prototype** is a key part of the manufacturing process because it helps to produce a higher **quality** finished product. You can use a prototype to test the appearance of your design, to check dimensions and to test jointing systems and the order of **assembly**. Prototyping in inexpensive materials helps prevent you from making costly mistakes later on.

Your task is to design a coin store to encourage children to save. It should be fun to use and store up to 100 coins. Make a prototype out of card, evaluate its visual appearance and modify the design if necessary.

a)   How large does the product need to be? Will it be upright or flat? What type of **system** will you use to get the money in and out?

b)   Sketch the final idea and work out how many component parts are needed.

c)   Cut out the parts in card and stick them together with masking tape. This prototype will test the appearance and **size** of your product and whether or not the pieces will fit together correctly.

d)   Modify the design if necessary. For example, is it fun to use? Could it be smaller?

e)   Once you are happy with your coin storage design, carefully take the pieces apart and write down all the dimensions. You will need to make allowances for final material thicknesses.

f)   Produce a full size **working drawing**, showing all the dimensions of the product.

g)   Work out and list the order of manufacture and assembly for the product (the order in which you will make the product pieces and then put them together).

Further work

- Choose appropriate materials to make your finished coin store.
- Decide what decoration you will use.
- Model the **joints**, fastenings and decoration before making up your product.

Name . . . .

Form/group . . . . **Date** . . . .

You can use **product analysis** to help you evaluate and compare the **quality** of a range of similar commercial products. This will enable you to develop ideas for setting up your own **quality control system**.

1   In a team, compare two similar products, such as kitchen clocks made by different **manufacturers**. One product should cost more than the other. For example, you could compare a plastic cased wall clock with a natural wood cased wall clock. It is likely that both will have a cover for the hands, one in clear plastic and the other in glass.

2   In the table below, compare the main design features and quality of the two products. Give each category marks out of 5 (5 = excellent, 1 = poor).

	Product 1	Product 2
Design features and quality of styling		
Material quality and suitability		
Quality of manufacture accuracy, finish, joints and fixings)		
Ease of use (e.g. setting and telling the time)		
Value for money		
Total score		

3   State which of the products is the most expensive, and why it costs more.

4   Which product offers the best value for money, and why?

5   Explain why each product is suitable for its **target market**.

Further work

- Write a **manufacturing specification** for the higher quality product.
- Explain how you could improve the quality of the lower quality product.

40   quality assurance

Name . . . .

Form/group . . . . **Date** . . . .

I   Using the table below, evaluate the last product you made for **quality of design and manufacture**. Explain how you could improve the **quality** of your product.

	Judging the quality of my product	How it could be improved
Quality of design		
Quality of manufacture		

2   Read the **quality assurance (QA)** procedures shown in the left hand column of the table below.

     a) In the middle column, tick the QA procedures you used to manufacture your product.

     b) In the right hand column, tick the QA procedures you could use to improve your product if it was manufactured in **high volume**.

     c) Choose two of the QA procedures you have ticked in the right hand column. Investigate how you could use them in your own product manufacture.

QA procedures	QA procedures used to manufacture my product	QA procedures I could use to improve the product if manufactured in high volume
Production plan		
Design specification		
Manufacturing specifications		
Product **costing**		
Work schedule		
Computer system (CAD/CAM)		
Quality control		
Inspection and fault finding		

Name . . . .

Form/group . . . . **Date** . . . .

Analyse the **quality** of a commercially manufactured product using the quality checks listed in the table below. For each quality check, write a sentence or two to describe your findings.

Quality check	Result of the quality check with reasons
Check for signs of scratches or product damage during or after manufacture	
Check if the fastenings, components or fixings are damaged or broken	
Check if the fastenings, components or fixings are faulty in any other way (e.g. corroding or rotting)	
Check that the **joints** are secure	
Check that all the fits are as intended (flush, flat, tight, accurate)	
Check that all the parts have been cut, machined or made correctly	
Check if any parts are misaligned or, if they move, whether they are stiff	
Check if there are any faults in surface detail including texture, **finish** or printing	
Check the product measurements match standard sizes. This might relate to seats and tables, etc.	
Check that any attached labels are easy to understand (e.g. for use and maintenance)	
Check if there are any attached labels related to the **safety** of the product in use	
Check that the product is safe for its intended users	

Understanding Industrial Practices: Resistant Materials Technology © Nelson Thornes 2004

42   quality control (2)

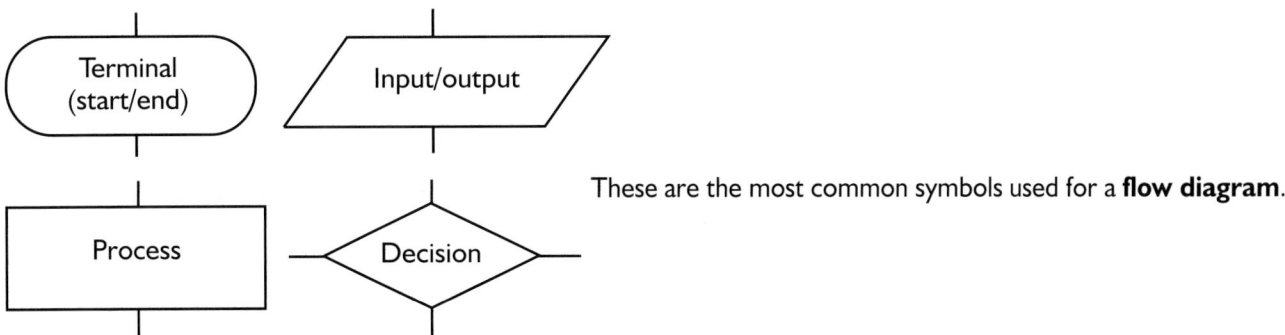

These are the most common symbols used for a **flow diagram**.

You will need to include **quality control (QC)** checks when you are designing and making your own product.

1 For your next product, investigate the **quality** checks, **tests** and **inspection** processes you need to undertake at the following stages:

- Incoming raw **materials**.
- **Production planning**.
- Processing.
- **Assembly**.
- **Finishing**.
- Final product.

2 Using the standard symbols shown above, draw a flow diagram to show where and how your quality checks will take place. Include **feedback** loops to show how feedback is incorporated into your QC system. As your work progresses, describe any changes you make to the design, processing, assembly or finish of your manufactured product to ensure that it will be of high quality.

3 Explain the difference between QC and **quality assurance (QA)**.

4 Explain how the use of QC ensures the manufacture of identical products in **high volume**.

Description of product: flat-pack seat		
Type of quality indicator	How and where used	Result of quality check
Variable	Measure the seat height	Sitting height is within tolerance (432 +/- 1 mm)
Variable	Measure the leg slot width	Leg slot is within tolerance (12 + 0.1 mm)
Attribute	Check for sharp or unfinished edges	Perfectly smooth finished edges
Attribute	Check paint finish	Very smooth surfaces with no thin or dull patches
Attribute	Check tightness and fit of hinges	Secure hinges

1   What is a **quality indicator**? Explain your answer.

2   Describe three ways that you could use a variable quality indicator to check the **quality** of a CD rack.

3   Describe three ways that you could use attributes to check the quality of a child's push-along toy.

4   Working in pairs, check the quality of a commercially manufactured product using quality indicators. Record your results in a table similar to the one below.

Description of product:		
Type of quality indicator	How and where used	Result of quality check

Understanding Industrial Practices: Resistant Materials Technology © Nelson Thornes 2004

44   quality of design

1 Read the information below about the **design theme**, **target market group**, and retail outlet.

Design theme

'Funky Monkey' desk accessories include a variety of desk tidy products, containers and stationery items. A wide range of contrasting resistant materials is used to make and decorate the products.

- Think quirky, moulded, wavy, wobbly shapes and fun styling.

- Think bold, vibrant **colours** such as red, blue, aqua, and white mixed with graphic darks such as charcoal, black and chocolate brown!

- Think shiny steel, copper or brass, natural or painted wood, brightly coloured plastic, 2D inlays or 3D surface decoration such as beads, shells or wobbly eyes.

Target market group

The target market is the 12–18 age group who are interested in novel, eye-catching products that are fun to use or show to their friends. This kind of accessory mixes fun with function.

Retail outlet

The retail outlets will be independent shops, such as 'Gadget Gifts', that sell unusual and eye-catching products.

---

2 Write a **design specification** for a range of desk accessories based on the theme 'Funky Monkey'. The accessories are to be used by the 12–18 age group either as **functional** items or for fun to brighten up their work area.

3 Produce some quick annotated sketches of your product ideas and evaluate your designs against your design specification.

4 Show your design ideas to potential users in your target market group. Using **feedback** from your users explain how you could improve the **quality** of your best design.

Further work

- Produce a **mood board** for your products.

- Use **computer-aided design (CAD)** software to develop and colour shapes and **styles**.

- Compare the **materials** and processes used for your own designs to similar products sold by the same types of retail outlets that you have designed for.

- Compare the functional features of your own designs to similar commercially available products.

---

45 quality of manufacture

In your own coursework project you should aim to design and manufacture a high **quality** product that meets your **manufacturing specification** and the needs of your users.

1 Write a manufacturing specification for your resistant materials product.

2 Explain what is meant by **tolerance** limits and state where they can be found.

3 Produce a **flow diagram** to show the order of manufacture and **assembly**.

    a) Using **feedback** loops show where you will check for quality at **critical control points** in your product's manufacture.

    b) List and explain the **quality indicators**, showing how you will check for quality.

4 Evaluate your product by comparing it with your manufacturing specification.

5 Obtain feedback about your product from users in your **target market group**.

6 Using feedback from your product users explain how the **quality of manufacture** of your product could be improved.

Understanding Industrial Practices: Resistant Materials Technology © Nelson Thornes 2004

Name . . . .

Form/group. . . . . **Date** . . . .

1 When designing and making your product it is important to reduce any risk to yourself and others. One way to make **safety** a priority is to think about safety with people, safety with **materials** and safety with equipment and machinery.

Complete the following check lists to help you design and make safely.

Safety with people

- Concentrate on your work to avoid accidents.
-
-
-

Safety with materials

- Take care when carrying long pieces of material in work spaces and through doors.
-
-
-

Safety with equipment, machinery

- Keep hands away from moving parts.
-
-
-

2 When designing and making your product you need to ensure that it is **fit for its purpose** and safe for the user. The table below provides safety points for **designers** of three types of play item. What other safety points can you think of for the push-along toy and the swing/seat?

Type of play item	Safety points for designers
Flipper ball game	The balls must be seen but must not be accessible, or be too small to swallow. Any see-through protective screen must be shatterproof. The product must have a non-toxic **finish**. It should not be possible to trap or damage fingers when using the game.
Cam action, push-along car	The product must not have sharp edges.
Swing/seat	The front of the seat should have a soft edge.

Remember to include safety criteria in your product **design specification**.

Name . . . .

Form/group . . . . **Date** . . . .

Work areas can be a potential source of danger.

1   Describe three **safety procedures** that a **manufacturer** can put in place to reduce risk and protect employees from danger in the workplace.

2   Describe three safety procedures that an employee can take to ensure that they work safely in the workplace.

3   Using the following table, make a list of do's and don'ts for using **materials** and equipment in your workshop.

Safety procedures in my workshop	
I must...	I must not...

Understanding Industrial Practices: Resistant Materials Technology © Nelson Thornes 2004

1. The **style** of many products depends on the **materials** and processes from which they are made, the **end-use** of the product and the **target market group**. Make a collection of images of well-designed products and investigate:

   - the style of each product. Is it **functional**, eco-friendly, trendy, streamlined or geometric, for example? Was the style of the product influenced by a **designer**, design group or architectural style?

   - are the styles suitable for fashion or functional products?

   - the type of materials and processes used. Are the materials hard, soft, expensive, shiny, dull, smooth, textured or easy to work? Is the product made from metal, plastic or wood? What types of manufacturing processes were used? For example, casting, injection moulding or machining.

   - the end-use of the product and the target market group. Was the product designed for adults or children? Where will the product be used? For example, on holiday, in the kitchen, in the garden or for a sporting activity.

2. Record your findings about each product using a table similar to the one below. Remember to include sketches or photos to show the proportions, shape, **size** and **colour** of each product. Make notes about the style, materials and processes used and the end-use of the product. Some examples are given in the table below. What kind of materials and processes would be suitable for the coffee set?

Product	Style	Materials and processes	End-use
	Influenced by the work of Le Corbusier Tall and thin Brown, yellow and red with a shiny finish	MDF and acrylic Cutting, slotting, piercing, machining, joining, gluing, heat bending, painting	Lights for use in the garden
	1970s and 80s 'Post-modernist' architectural style Shiny brown metal with green, grey, gold and black detail		Coffee set

Further work

Choose a **design theme** for your next product, such as 'nursery rhymes'. This kind of theme can be used for adult or children's products.

- Create a **mood board** to develop your design theme for a small child's furniture product, using your investigation into well-designed products to help you.

- Use sketches, notes and photos to help develop your mood board ideas.

Name . . . .

Form/group . . . . **Date** . . . .

Computer numerically controlled (CNC) machines enable the fast production of manufactured component parts. **Production systems** using CNC machines consist of three sections:

- a CNC machine
- a computer with a screen
- **computer-aided design (CAD)**.

This **system** is represented by the block diagram below:

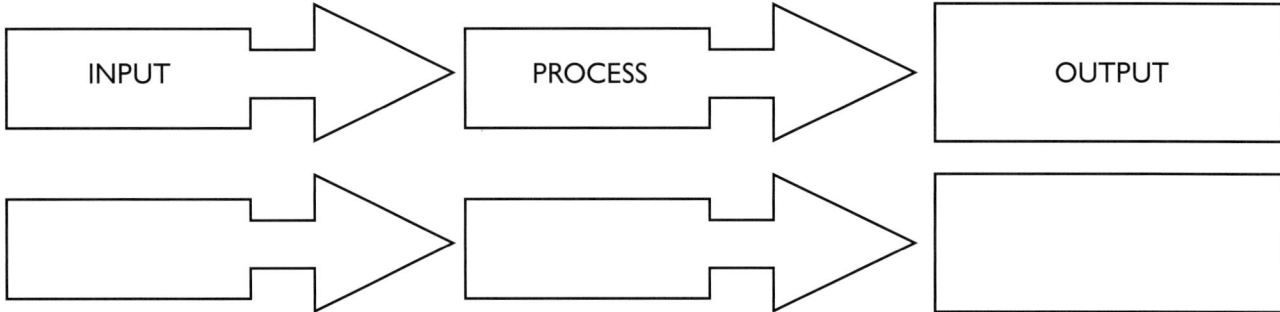

1 Put the three sections listed above in the correct place in the block diagram.

2 Name the kind of system represented in this block diagram.

3 Using a similar block diagram, show and explain how you could improve this system.

You can plan a project using a simple open loop **systems** approach. This can help to organise your tasks.

1 Make a list of all the raw **materials**, components, fixings and fastenings you used to make your last product. Include all the processes you used, such as making **joints**, machining, cutting, forming, **assembly** and **finishing** processes. List all your made component parts, **sub-assemblies** or completed outcomes.

2 Draw a table with three columns headed Input, Process and Output. You could use a table similar to the one shown below. Fill the 'Input' column with the appropriate raw materials, fixings and fastenings, the 'Process' column with manufacturing processes, and the 'Output' column with your finished component parts, sub-assemblies and outcomes from your lists.

Input	Process	Output

3 Using your Input, Process and Output lists, design a simple open loop system using a block diagram template similar to the one used in worksheet 49. This block diagram should show how you achieved an outcome such as a component part or sub-assembly. Alternatively, you could show how the finished component parts become the Inputs, the assembly techniques become the Process and the final product becomes the Output.

4 Using a block diagram template similar to the one shown below, organise your Input and Process lists into two vertical **flow diagrams**. In each flow diagram show the sequence of the activities or processes you undertook to make the component, sub-assembly or final product. By doing this you will have created sub-systems within your large system. Remember to write your product outcome in the Output box.

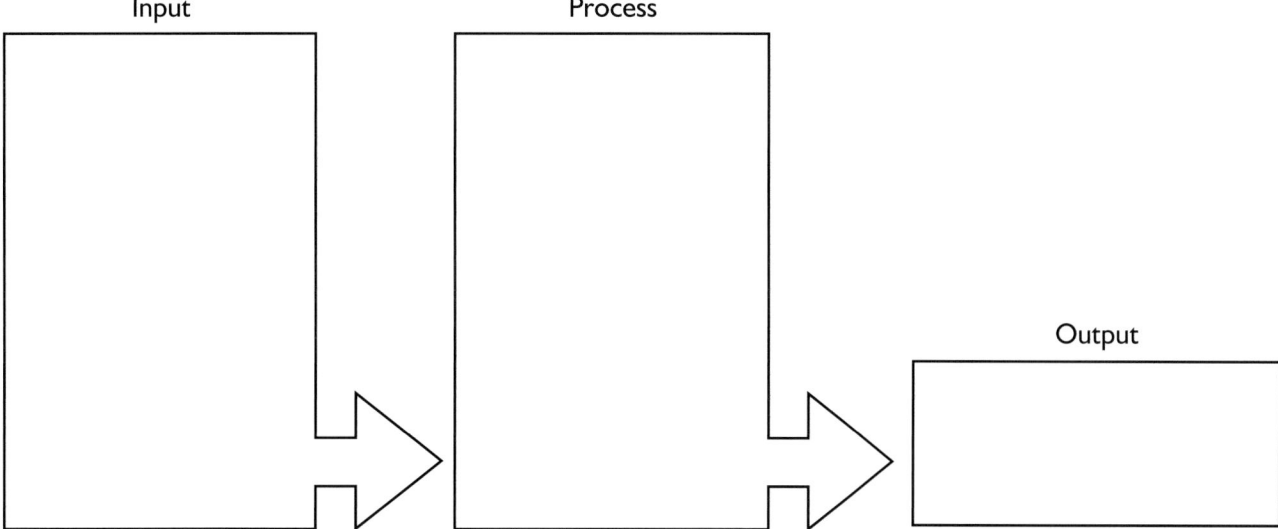

51   using a systems approach (2)

You can plan a project using a closed loop **systems** approach. This can help to organise your tasks and enable you to use **feedback** to monitor the **quality** of your product for a better result.

1   Undertake Tasks 1–4 in worksheet 50.

2   You are now ready to introduce quality checks into your system. Choose some, or all, of the following to check the quality of your product:

Check:
- **materials** and equipment
- against your **design specification**
- against your **manufacturing specification**
- the sizes and critical **tolerances** from your **working drawings**
- for **safety** in manufacturing
- **assembly** quality.

All of the tests or checks you have chosen can provide feedback to your quality system helping you to produce a higher quality product.

3   Using a block diagram similar to the one shown below (and the information from your sub-systems block diagrams produced in task 4 on worksheet 50), produce a new sub-systems diagram that includes feedback loops to show how and where your checks monitor and improve the quality of your product. (Note that you will need more space than available in the illustration below.)

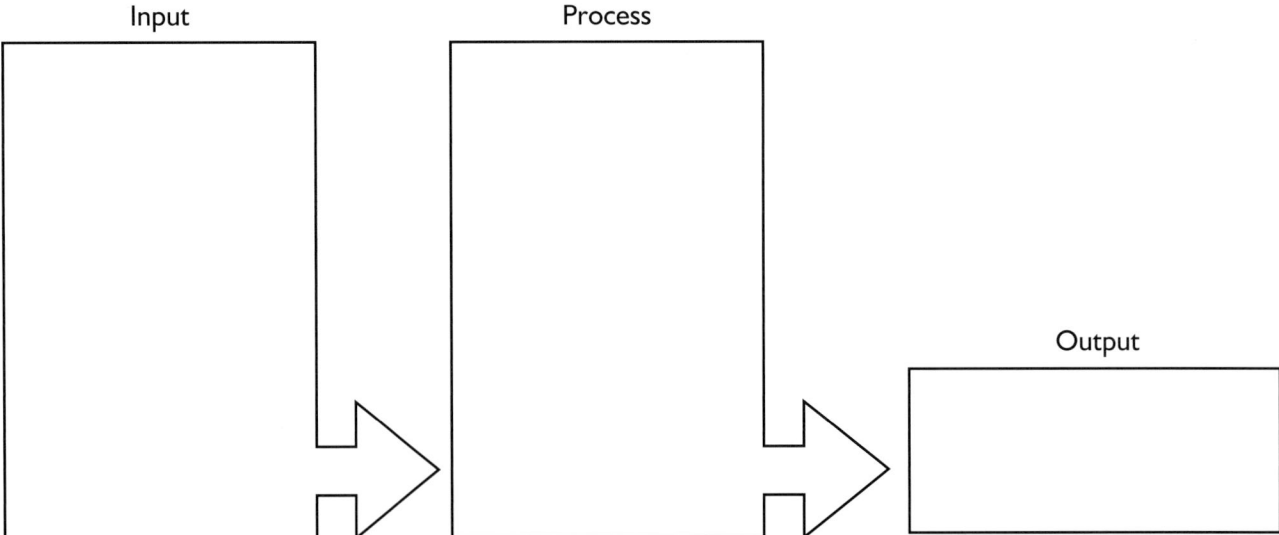

Further work

Plan your next project using a systems approach. You will be working in a similar way to industry. You will also help yourself to work more efficiently and produce a high quality manufactured product.

Understanding Industrial Practices: Resistant Materials Technology © Nelson Thornes 2004

Identifying the **needs or wants** of users in your **target market group** enables you to develop a product that meets existing market needs.

Focus your target market group research by collecting information related to:

- **style** trends
- cultural, environmental or designer influences
- social influences, such as travel or music
- a specific type of retail outlet, e.g. a DIY store, toy shop
- analysis of a similar commercial product.

1. Research the buying behaviour, taste and lifestyle of your target market group.

   Find out about:

   - the gender and age range of your target market group

   - taste: what are the likes and dislikes of your target market group in the product area you are considering?

   - lifestyle: what type of leisure activities and hobbies does your target market group like? Are they hikers, music lovers, gardeners, sporty or DIY enthusiasts?

   - cost needs: how much money do users in your target market have to spend on essential or luxury items? What level of product **quality** are they looking for? What type of retail outlets do they prefer to use (include shops and mail-order catalogues)?

   - preferred **brands**: what kind of brands do users in your target market buy? Are they prepared to pay more for branded products? Do they buy 'own-brands'? What is the brand image?

2. Hints for writing a target market group research questionnaire:

   - Decide what information you need to find out and how many people to ask.

   - Use a word processor to create a professional look to your questionnaire.

   - Phrase your research questions to get yes or no answers or use multiple choice questions with tick boxes similar to the example below.

Questionnaire – Portable CD rack for home use Please tick the appropriate box(es)				
How many CDs do you play regularly?	0	1–10	11–20	20+
Are you male or female?	Male		Female	
What is your age group?	Under 12	12–18	18–30	Over 30
Which shop's music accessory product styles do you prefer?	Department store	Computer store	Catalogue	Music store
What type of music do you prefer?	Pop	Rock	Jazz	Classical
How do you store your CDs now?	Box	Drawer	Shelf	Rack
What price would you pay for a CD rack?	£5–10	£10–15	£15–20	£20+
What is the most important feature of	Function	Style	Image	Cost

3. Analyse the results of your research and include a summary in your coursework folder. If you use a spreadsheet for this analysis you will demonstrate your use of **Information and Communication Technology (ICT)**.

You should only test the **materials**, processes and **constructions** that you intend to use in your coursework product. These tests should be specifically aimed at the **properties** and characteristics the end product will need. You should aim to keep **testing** to an essential minimum.

Testing materials

Try using the following simple method of testing materials:

- Look up the expected properties of your materials in a textbook.

- Check that the properties are appropriate for the product you are designing.

- Test the materials and compare your results with the expected properties.

- Write up the results of your testing and explain why the chosen materials are suitable for your product.

Remember that testing needs to be carried out under controlled conditions. You must test each material in exactly the same way for it to be a fair test. For example:

- If you are testing shelf materials for flex each material tested must be of the same length and cross-section and you must use the same weights in the same place each time.

- If you are testing materials for glue strength you must use the same size gluing surface area each time.

Testing processes

Test which material processes are easiest to use that meet the requirements of your **specifications**. For example, it is usually much easier to bend or form plastic sheet than it is to join it to make a 3D form.

Try using the following simple method for testing processes:

- Look up processes you could use for your product and decide which is the most appropriate and easiest to use. For each process you test, use the same materials you intend to use for your product.

- Write up the results of your testing and explain why the chosen processes are suitable for your product.

Testing constructions

If your product has component parts joined together it is a good idea to test sample **joints** before final construction. This will enable you to see if you can make the joints and if they are strong enough for your product.

Try using the following simple method for testing constructions:

- Look up methods for joining together the component parts of your product.

- Test which is the easiest, strongest and best looking by making sample joints using your chosen methods. You may decide to use mechanical fixings, traditional joints or adhesives. Note that the easiest joint may not be the strongest or best looking.

- Write up the results of your testing and explain why the chosen constructions are suitable for your product.

Continued on worksheet 54

Also see worksheet 53.

Testing finish

You need to test suitable **finishes** for your chosen **material**, particularly if the product has to be used in a difficult environment like outside. Finishes should be safe for children, especially if they put the product in their mouth.

You could try using the following simple method of testing finishes:

- Look up information about suitable finishes for your chosen materials. Draw up a table to explain which would be the best finish to use and why. Remember that some woods can be left with a natural sanded finish, depending on use. Most plastics are self finished. Metals can be painted, dip coated or plated.

- Depending on the **end-use** of your product you may need to test finish for: water resistance, **colour**, feel, appearance or permanence.

Testing manufacturing quality

Test manufacturing **quality** at **critical control points** by using the **quality indicators** outlined in your **quality control flow diagram**.

You should aim to produce an end product that meets all the criteria in your **specifications**.

Testing the product in use

Testing the product in use is essential to find out if the product is successful.

You could try using the following method for testing the product in use:

- Run a **user trial** of your product in its intended environment to see how well it meets your specifications.

- Ask users in your **target market group** if they like or dislike your product. Include questions about the way that the product functions and its appearance. For example, if the product was for a young child is it durable and does it keep the child's attention?

- Ask the target market group if they would buy your product at the selling price you set.

Values issues have an important influence on our lives and on the products that we use. **Designers** have to carefully consider the values, **needs and wants** of the **target market group** at which the product is aimed.

You will need to make use of social, cultural or environmental influences when you develop your own product ideas. It will help you to think about values issues if you ask some of the following questions when you undertake product analysis:

About the product

- Who is the target market group for this product? What are their needs?
- Why would the target market buy this product?
- Is it a trendy item or a **functional** product?
- Would you like to own or use the product? Why?

About the need for the product

- Is the product really needed? Is it a 'must have'?
- How well does the product meet the needs of the target market?
- Who benefits from the manufacture of the product? How will they benefit?

The design of the product

- How is the product special or different from existing products? Is it a **branded** product?
- What kind of image does the product give?
- Is there a choice of **style** or **colour**?
- What kind of values influenced the design of the product? Cultural, social or environmental?

The product manufacture

- What kind of **materials** and processes are used to make the product, and why?
- Where do the materials come from? Will they run out?
- What impact does the use of these materials and processes have on people or the environment?
- What happens to any waste produced during manufacture?
- What skills are needed to make the product?
- Where is the product made (at home or overseas)?
- What are the working conditions like in the place of manufacture (safe, clean, good facilities)?

Advertising and marketing the product

- How is the product advertised and packaged?
- How and where is the product sold?
- Is the product sold at a suitable cost for the target market?
- How does the cost compare with other similar products?

Use and disposal of the product

- What is the product used for?
- Is the product safe for the user and the environment?
- How long is the product expected to last? How will it be disposed of?
- Could the product be made to last longer and if so how?
- How easily can the product be recycled? Who pays for the cost of **recycling**?

Using a table similar to the one shown below, write a **work schedule** for manufacturing the component parts and **assembly** of your coursework product.

1 List the component parts.

2 Put the processes in order and estimate how much time each one will take.

3 Include the following information in your work schedule:

- A title and date.
- Component parts.
- Processes.
- **Machines** used.
- Time taken for each process.
- Total process time.
- Where to find dimensions and critical **tolerances**.

Write the components parts in the work schedule in the correct order they need to be processed.

Note that dimensions and critical tolerances for the component parts are given on the **working drawings**.

In the work schedule shown below the timings are for a **one-off** jewellery box made by a craftsman in a school workshop. They do not include machine set-up times.

Work schedule: Jewellery box See working drawing for dimensions and critical tolerances				Date: 18 Nov	
Order	No.	Component part	Process/assembly	Machine	Time (minutes)
1	1	Box sides (uncut lengths)	Groove bottom edge for base	Router	5
2	4	Box sides	Cut to length	Band saw	4
3	4	Box corners	Comb joints (cut and fit)	Band saw & jig	60
4	1	Base	Cut to size	Band saw	5
5		Box sides & base	Sand to finish	Belt sander	5
6		Box	Assemble joints and base and glue	By hand	10
7	1	Top	Cut to size	Band saw	10
		Box	Finish, sand and apply wax		10
				Total process time =	105

Accurate **working drawings** are a very important part of the **manufacturing specification**. They provide essential details about the design and manufacture of a product including dimensions and critical **tolerances**.

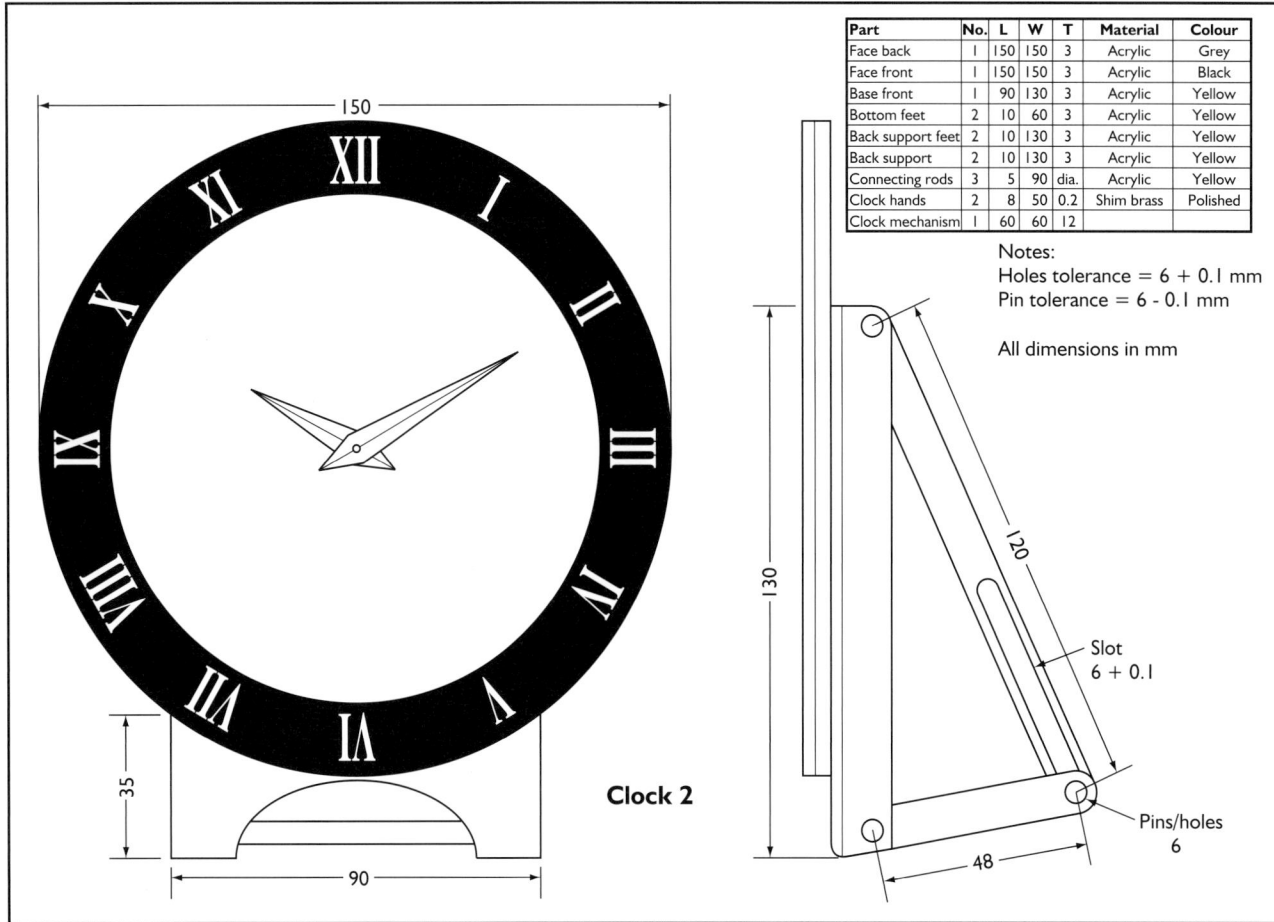

Part	No.	L	W	T	Material	Colour
Face back	1	150	150	3	Acrylic	Grey
Face front	1	150	150	3	Acrylic	Black
Base front	1	90	130	3	Acrylic	Yellow
Bottom feet	2	10	60	3	Acrylic	Yellow
Back support feet	2	10	130	3	Acrylic	Yellow
Back support	2	10	130	3	Acrylic	Yellow
Connecting rods	3	5	90	dia.	Acrylic	Yellow
Clock hands	2	8	50	0.2	Shim brass	Polished
Clock mechanism	1	60	60	12		

Notes:
Holes tolerance = 6 + 0.1 mm
Pin tolerance = 6 - 0.1 mm

All dimensions in mm

Clock 2

1 Copy the above diagram to exact dimensions using a **computer-aided design (CAD)** system. You will need to set your paper size to A3. You may have to use dots for hour indicators if you don't know how to use numerals. Other details such as the curved cuts may also have to be simplified.

2 Insert all the sizes on your drawing.

3 Sketch a clock design of your own and then produce a full size CAD view from your sketch details.

4 Produce a **materials** list and critical tolerances for your own design.

Further work

1 Produce working drawings for your own project.

2 Produce a materials list and critical tolerances for your own project.

Notes

Notes

Notes

Notes